THE KEEPERS OF WATER AND EARTH

THE KEEPERS OF WATER AND EARTH

Mexican Rural Social Organization and Irrigation

by Kjell I. Enge and Scott Whiteford

Foreword by Robert C. Hunt

UNIVERSITY OF TEXAS PRESS AUSTIN

First edition, 1989

Requests for permission to reproduce material from this work should be sent to:
Permissions
University of Texas Press
Box 7819
Austin, Texas 78713-7819
utpress.utexas.edu/about/book-permissions

Library of Congress Cataloging-in-Publication Data

Enge, Kjell.
The keepers of water and earth : Mexican rural social organization and irrigation / by Kjell I. Enge and Scott Whiteford.—1st ed.
p. cm.
Bibliography: p.
Includes index.
ISBN 978-0-292-75397-6
1. Irrigation—Economic aspects—Mexico—Tehuacán River Valley. 2. Peasantry—Mexico—Tehuacán River Valley. 3. Tehuacán River Valley (Mexico)—Rural conditions. I. Whiteford, Scott, 1942–
II. Title.
HD1741.M62T44 1989
307′2′097248—dc19 88-35042
CIP

♾ The paper used in this publication meets the minimum requirements of American National Standard for Information Sciences—Permanence of Paper for Printed Library Materials, ANSI Z39.48-1984.

First paperback printing, 2013

CONTENTS

FIGURES

TABLES

The hot sun baked the dusty soil, drawing out the last traces of moisture. Stretching in uneven rows, the corn was turning yellow-gray as it began to wilt. For centuries, the field had been carefully tended by the inhabitants of the valley, who had constructed a series of canals to guide irrigation water from distant springs. Their ancestors, robbed of the water by the Spaniards, had regained it after prolonged conflict and at great cost. New sources of water were later tapped through a tunnel system of chain wells, constructed at great economic and physical sacrifice. Yet today, the fields were left to dry because the owners of the land could not afford the water to irrigate their crops. In another field, located only a few hundred meters away, well-watered grapevines and other commercial crops were being carefully tended by hired laborers. The fruit would soon be harvested by seasonal laborers and exported. The perpetual struggle for water, the key to life in the valley, had entered a new phase. The valley that had witnessed the development of one of the earliest irrigation systems in the New World was now being transformed by new technology and socioeconomic organization. Like other changes that had marked the valley's history, the ecological and social fabric that made life possible for a growing population was being transformed.

FOREWORD

IRRIGATION HAS been terribly important to the inhabitants of valleys in relatively arid zones for millennia. In extreme cases it is what makes settled human occupation of a valley possible. Many cases are not quite so extreme, and here irrigation makes possible the existence of cities and of large populations. For millennia humans have known that irrigation water gives life, and that the human investment in the irrigation works, and organizations, has paid off handsomely.

Most of us who live in relatively humid urban zones pay very little attention to the water that keeps us healthy and fed. In spite of the fact that at least 50 percent of what humans grow is under irrigation, in spite of the fact that irrigated hectarage has more than doubled in the past forty years, in spite of the fact that the Green Revolution, which has bought for the human population a generation of respite from famine, has depended utterly on irrigation facilities, there is remarkably little interest in irrigation either as a technology, or as a locus for important social organization.

V. Gordon Childe offered the idea that irrigation made possible the surplus without which the urban revolution could not have happened (1936). Although largely true, most scholars do not think very much about the matter. By far the most prominent name associated with irrigation is Karl Wittfogel, who is said to have argued that irrigation was causally linked with Oriental Despotism (1957). There has been no definitive test of Wittfogel's major hypothesis, which concentrated not on irrigation alone, but on hydraulic works (Mitchell 1973).

There have been precious few attempts to actually study the way that irrigation is organized. Most social and cultural anthropologists in the past headed up into the hills, where most of the "traditional" peoples lived. In most of the world there are few irrigation facilities high up in the hills (but see Conklin 1980 and Guillet 1987). Rather, most irrigation systems are down in the valleys, for obvious geomorphological reasons.

The general ignorance of these irrigation systems has often unfortunately extended even to those who study people who live by irrigation down in the valleys. There is a most curious blindness about irrigation on the part of most social scientists.

Enge and Whiteford's *The Keepers of Water and Earth* is a happy exception to the general rule. They have worked in the Tehuacán Valley, which has been civilized, and irrigated, for over two millennia (see Woodbury and Neely 1972). They have looked not only at the irrigation systems, but also at how important they are, and how they articulate with the rest of the region. This is one of a handful of studies with this kind of focus, and is most welcome.

There is however another reason for welcoming this study. Most of the gravity irrigation systems of this world have captured surface flows. There is another technology, however, that of the qanat (the digging of horizontal wells to tap into an aquifer many kilometers away; see Wulff 1968). Jeffrey A. Gritzner (1977) has detailed how this particular technology is distributed in Ancient Near Eastern space and time, and John C. Wilkinson has studied its role in Oman (1977). It diffused eastward into China, and westward along the southern shore of the Mediterranean. Iran is where the most intensive utilization has taken place (English 1966; Spooner 1974[a], 1974[b]).

The water capture technology in use in the Tehuacán Valley includes qanats, the only well-known instance of this technology outside the Old World. And it so happens that Enge and Whiteford have more information on the social organization of these qanats than is currently in print in English for any part of the world. All students of irrigation social organization will be grateful for this study.

ROBERT C. HUNT

PREFACE

THIS BOOK, like most books, represents only a part of the research findings of the authors' work in the Tehuacán Valley; our names are listed alphabetically and do not reflect any form of seniority.

Kjell Enge began his research in the valley in 1970 and has returned regularly since then. He was based in the town of Altepexi. His long-term research has given him the opportunity to observe significant changes in the valley and to explore their causes and implications. Using an actor-focused conceptual framework, he has examined the competition over access to resources with special emphasis on actor strategies and tactics within the organizational context. He had unique access to irrigation association membership and budget records, which allowed him to document conflict mediation and membership changes. Kjell wishes to express his thanks and appreciation to the members of the water associations who so freely and generously provided the data on local history, problems, and operating procedures, the engineers of the Papaloapan Commission office in Tehuacán, and the people of Altepexi and the neighboring *municipios* who made his research possible. In particular, he would like to recognize the tireless work and efforts of Pilar Martínez, his wife, who translated old handwritten documents, interviewed informants, and, in the process, became an experienced ethnographer; and Eva and Robert Hunt, who first introduced him to Latin America and the Tehuacán Valley.

Kjell's first visit to the Tehuacán Valley was made possible by a summer grant from the Wenner-Gren Foundation for Anthropological Research; an extended stay during 1972 and 1973 was financed by a grant-in-aid from the Wenner-Gren Foundation. Subsequent visits during the summers of 1974 to 1978 were made while teaching in

Costa Rica and Guatemala on a Latin American Teaching Fellowship from the Fletcher School of Law and Diplomacy at Tufts University. The summer of 1981 was underwritten by a research grant from Baylor University, and the following two summers were part of a summer field school in ethnology directed by him and funded by Baylor University. Further support for data analysis and manuscript preparation was provided by research grants and logistical support from Dickinson College.

The research by Scott Whiteford was part of a project carried out between 1975 and 1977. Scott and Weegee Whiteford lived in the community of San Gabriel Chilac. Weegee, a professional writer in her own right, brought a special insight and perspective to the research. Both Scott and Weegee want to express their appreciation to the many people of the valley who took them into their homes and shared their lives with them. For the Whitefords, whose son, Ian, was born in Tehuacán, the valley will always hold a special place in their lives. Scott has returned to the valley several times, most recently in 1988.

The Tehuacán research was a team project supported by the Centro de Investigaciones Superiores of the Instituto Nacional de Antropología e Historia. Other members of the team were Consuelo Ocampo, who worked in San José Miahuatlán; Sergio Quesada, who lived in Calipán; and Luis Emilio Henao, who focused on Ajalpan. Their research examined the expansion of capitalist agriculture in the valley and its impact on household organization and communities. Understanding the evolution of patterns of control, means of production, and the articulation between different modes of production was a critical dimension of Scott's research. He is indebted to all members of the research team mentioned above, as well as to Catalina Rodríguez. All made lasting contributions as scholars and as friends. A special debt of appreciation is due the late Angel Palerm, whose standards of scholarly excellence and theoretical creativity provided inspiration and friendly criticism.

A fellowship from the National Endowment for the Humanities made it possible for Scott to spend a stimulating period at the School of American Research, where he completed his work on the manuscript. He is grateful for the challenging discussion generated by fellow scholars Garrick Bailey, Bob Canfield, Christine Rudicoff, and Flora Clancy, and to the president of the school, Doug Schwartz. Both authors would like to thank the reviewers of the manuscript and Helen Pollard for their constructive comments.

The collaborative book emerged after the authors discovered, much to their delight, that each was working on a manuscript on similar issues in the same region. Because of mutual interests in the critical role that control of water resources played in the social organization of the valley and the unique opportunity the situation presented for the examination of broader theoretical problems, they decided to combine efforts. The experience has been both stimulating and challenging. For helping the manuscript to become a reality, we would both like to thank Weegee Whiteford for her excellent editorial work and Elaine Mellen for making endless corrections and revisions, typing tables, and tending to all the details.

THE KEEPERS OF WATER AND EARTH

1. MEXICAN RURAL DEVELOPMENT AND IRRIGATION

AS THE world's population expands, agrarian producers are facing growing pressure to increase food production. At the same time agricultural exports are expected to generate foreign revenue for natural industrialization and military growth. In arid regions, massive amounts of money are being invested in irrigation infrastructure and government bureaucracies in an effort to convert deserts into croplands. By the mid-1980s irrigation covered 18 percent of world cultivated land, producing one-third of the world's food (Rangeley 1987:30). Investment by international agencies in river basin development in addition to thousands of national projects guarantees the expansion of irrigated regions worldwide.

In Latin America irrigation dates back to the forerunners of the elaborate irrigation systems of the Incas and the Aztecs. Today, irrigation provides the basis for agriculture in some of the most productive regions ranging from Mendoza, Argentina, to Northwest Mexico.

Of all the Latin American nations, Mexico is the most dependent on irrigation for its agriculture. More than 45 percent of the country's commercial crop production is raised on irrigated land (Yates 1981:64). The Mexican government is under tremendous pressure to increase agricultural production. Faced with a rapidly growing population and the second-largest foreign debt in the world, and fearing political blackmail should the nation become dependent on imported food, the government has allocated an increasing amount of fiscal and managerial resources to agriculture. Early in the twentieth century, Mexico nationalized water resources and began to develop a sophisticated bureaucratic structure to implement water management and agricultural policy.

Irrigation is critical to Mexico's future because the nation is not well endowed with rainfall or natural water resources. Forty-three percent of the arable land is impossible to develop for agriculture without irrigation, and 34 percent has only seasonal rainfall. This leaves only 23 percent with adequate natural water. Unfortunately, 7 percent of Mexico (the southeast section of the country) has 40 percent of the national water resources. Only 12 percent of the water resources are in the Central Plateau, which has 60 percent of the population and constitutes 51 percent of the land mass. Athough 85 percent of the water resources are located below the elevation of five hundred meters, 70 percent of the population and 80 percent of the industry are located above five hundred meters (Scott 1982:311). In 1970, 95 percent of the water used in Mexico was allocated to agriculture. The water requirements in this sector of the economy are expected to more than double by the year 2000. At the same time, industrial and domestic use of water is expected to triple (Secretaría de Recursos Hidráulicos 1975).

The irrigation systems that currently exist throughout Mexico are plagued with serious problems of salinization and silting. The most significant of these systems is in the Mexicali Valley, where saline water carried by the Colorado River from the United States has destroyed one of the country's most productive agricultural regions.

Many of Mexico's major agricultural regions are in need of major rehabilitation. "At a conservative estimate, something like 2.5 million acres will need investment in rehabilitation before 1990" (Yates 1981). These figures indicate the scale of investment necessary to develop and maintain irrigated agriculture in Mexico and why there has been such a demand for government participation in the financing and construction of irrigation projects. At the same time, the figures point to the necessity for careful management of water resources and explain why government officials want to be involved in the management process after the investment is made. Yet increased government involvement has not resolved many of the problems. For example, maintenance of the systems is often neglected by the users, who have the responsibility for upkeep, leading to expensive water loss. District committees in charge of distributing water may be highly political. Although in some cases water is distributed equitably, in others it is allocated according to acreage sown, or directly to local bosses and their friends (Yates 1981:81).

The allocation process clearly illustrates an important dimension of irrigation systems that is often forgotten by engineers: irrigation systems are parts of more complex social systems. They are managed by people who share customs, traditions, and laws that are part of a common cultural system. The transformation of the natural environment to make cultivation possible is one aspect of the production process and must be seen in relation to the control of means of production, the technology, the social relations of the people involved, and other dimensions of the sociocultural system, especially the state.

One of the most important processes in twentieth-century Latin America has been the expansion of state power and the consequent national control of administrative activities. In many countries, including Mexico, the mechanisms for increasing state control have consisted of the nationalization of water, minerals, petroleum deposits, or land. In Mexico the government has extended its control over irrigation systems to most parts of the country.

The state has taken an increasingly active role in implementing national programs in the name of rural development. Although many of these policies and programs are supposed to be distributed equally throughout a given country, distribution is more often than not uneven and disproportionate, both within and between regions. A classic case is the Mexican agrarian reform, implemented in different ways at various times in diverse regions of the country. The reason for uneven policy implementation is, we hypothesize, embedded in the nature of local social organization and how it links or articulates with regional and national centers of power.

The social consequences of government control over basic resources such as water and land are tremendous. Once the government controls access to land or water, it increases its power, especially that of bureaucratic officials. This can lead to greater or reduced access to the basic means of production, depending on the nature of the government and the economic system. The power relationship definitely allows the government to dictate the conditions that must be fulfilled if people want access to the resources.

In the Mexican case the government also controls another key resource—credit. If people want access to these resources, they often must plant what the government wants, use the techniques prescribed by the government, and respond to other requirements. For example, in many areas near sugar mills, the government requires the peasants to grow sugarcane, even though they could make more money growing other crops. In order to institute particular programs, the government either captures and transforms traditional institutions (institutional penetration) or creates new institutions (structural penetration) to implement the goals of the government leaders (Corbett and Whiteford 1983). In Mexico, both forms of penetration are found in regions of irrigated agriculture. Where new irrigation systems have been created, the entire institutional structure is relatively new; in regions where irrigation has a long history, the government co-opts or changes traditional institutions as the management of the systems becomes centralized.

In this book we examine the process of the expansion of state power in a region that has a long history of strong local control over agricultural production punctuated by periodic change and conflict with outside interests. The documented history of the Tehuacán Valley, located in the state of Puebla, goes back to prehistoric times of migrant hunters and gatherers who later settled, developing irrigation agriculture. The settlements evolved into highly sophisticated nascent city-states prior to their incorporation into the expansionist Aztec state. Shortly after being incorporated into the indirect tribute system of the Aztecs, the valley people became subjects of the Spanish Crown for the next three hundred years. Independence, revolution, and incorporation into the modern world system brought even more pressures on and changes in the institutional structure of the Tehuacán Valley. Events since the Revolution have been variations on an oft-repeated theme: competition for productive resources that pits local interests against outsiders.

Our focus is on the twentieth century, although we recognize the importance of historical events and patterns going back to the earliest times. In order to better understand the current situation, we present data on two similar communities in the valley, each affected by the postrevolutionary redistribution of land and water. To examine the long-term consequences of reform for communities and the region, we ask why the implementation of uniform national agrarian policies has had such strikingly different effects on two neighboring communities. Such questions are critical because they lead us to examine the nature of linkages between macro and micro levels and the articulation of community social structure to national centers of power. Equally important, the questions help focus the analysis on how individual households compete for and have gained access to resources both within and outside of their own localities.

More specifically, this book is about the people of the Tehuacán Valley, whose agriculture is based on a complex irrigation system that was developed without aid from the government and is maintained with a high degree of autonomy. Although nominally incorporated in the Papaloapan district, water resources are privately owned, freely bought and sold, and, in one community, exchanged on an open futures market. The following pages describe the evolution of the agricultural system in the valley and focus on the sociocultural and economic relationships that generated the system, as well as on those that were created by the structure. In order to put the study in a broader context, it is important to sketch the process by which management of water resources in Mexico developed.

AGRICULTURE AND IRRIGATION IN MEXICO

In Mexico the control and management of irrigation has a long history. Beginning as early as 1500 B.C. the people in both the highlands and the humid tropics developed a sophisticated system of water control that included elaborate canal networks, dikes, dams, and aqueducts. After the Conquest, the Spanish, with their own extensive experience with irrigation, brought new irrigation technologies to the New World; from Zacatecas to Oaxaca they dammed rivers, tapped groundwater, and built extensive canals and monumental aqueducts.

From the time of Mexico's independence in 1821 until the despotic reign of Porfirio Díaz began in 1880, political instability in the countryside impeded the construction of large irrigation systems. Expansion of irrigation began under the dictatorship of Díaz, who brought a period of forced stability. Díaz gave concessions to companies, many of them foreign, to develop agricultural lands that required irrigation. The government underwrote much of the construction expense of these systems, which were concentrated in Morelos and the Valley of Mexico (Reyes Osorio et al. 1974:864).

During this period the native agricultural sector was neglected, if not abused, by the national government. Standards of living in the countryside were very low, food prices skyrocketed, and landowners extended their power at the expense of the peasantry. These abysmal conditions played a major role in sparking the Revolution of 1910.

The overthrow of Porfirio Díaz and the subsequent struggle for power and reform left the countryside torn by conflicting parties. The unstable situation made the expansion of irrigation systems impossible. Although land reform was the rallying cry of the Revolution, the importance of water was not forgotten by the framers of the Revolutionary Constitution of 1917, who included as Article 17 the nationalization of water. Various loopholes were left to protect the current system by allowing the government to transfer control of water resources to the private sector when necessary. Several years passed before government agencies were created to supervise water resources. In 1921 the Department of Agrarian Concerns and Colonization (Departamento de Asuntos Agrarios y Colonización—DAAC) was created as part of the Ministry of Agriculture (Secretaría de Agricultura).

The irrigation law of 1926 was pushed through by President Plutarco Elías Calles. This law nationalized private irrigation systems, legislated the construction of new irrigation systems, and established the

National Irrigation Commission (Comisión Nacional de Irrigación—CNI). A fund was created to finance construction of irrigation works, and a method was developed to incorporate ejido members' payments to help finance part of the construction (Orive Alba 1960). Despite its growth, the CNI was not allowed to develop broad-based integrated rural development programs, but was restricted to the development of irrigation systems. Starting in 1930, it created a series of irrigation districts. The irrigation law established the payment schedules for water that formed the pattern for years to come.

The push to expand irrigation resources during this period was sparked by Mexico's population explosion.

> The Mexican government faced the problem of feeding a population which increased by 78 percent between 1920 and 1950. The value of agricultural production during this period increased by about 54 percent, and it is doubtful that even this gain could have been achieved without the irrigation program. Had the benefits brought about by the CNI been absent, the nature of Mexican social and political life would have been radically different. (Greenberg 1970:19)

From its foundation the CNI was highly political. The creation of new irrigation projects was politically motivated, as the central government, located in Mexico City, maneuvered to tighten its control over regions that had resisted national control. "Projects [irrigation] in Aguascalientes, Tamaulipas, and Río Tula were all selected because of political pressures" (Greenberg 1970:13). Projects along the border with the United States were initiated to strengthen Mexico's position in bilateral conflicts over water and even territorial sovereignty. Between 1941 and 1946, 15 percent of the federal budget was allocated to irrigation projects—90 percent of the total funds allocated to agriculture. When the Ministry of Water Resources (Secretaría de Recursos Hidráulicos—SRH) was created in 1946, it had the second-largest budget of any government ministry. Between 1940 and 1946 alone, cropland covered by government-financed irrigation tripled in size (Hansen 1971:44).

Despite its large budget, the organization was still caught up in competition with other agencies that were also charged with the mission of rural development. The trend toward increased state control was slowed by a law passed in 1947 that allowed for large private holdings of irrigated land. The new law sanctioned the expansion of private agriculture in irrigation districts, as well as the transfer of water rights when land was confiscated by the state for redistribution. Water could be sold separately from the land in these cases. This law allowed the private sector greater opportunity to control irrigated land and water resources. The percentage of the federal budget invested in irrigation fell during the 1950s, but investment during the 1950s and 1960s expanded the acreage of irrigated land at a rate of 4.9 percent a year (Hansen 1971:44).

Today in Mexico, irrigation districts are one of the most important units of administration of agriculture and water resources. The districts are run by a committee of the CNI that is responsible for decisions about crops to be planted, distribution of water, types of fertilizers and insecticides to be used by farmers, and maintenance of the system. Yet the SRH and the Ministry of Agriculture generally set the policies and have veto power over committee decisions. The committees are thus relegated to a limited role (Greenberg 1970:29).

During the 1970s the SRH continued to expand, creating a reputation for efficiency and professional administration. Yet it was always in competition with the Ministry of Agriculture for financing, and in many cases the two ministries duplicated services. President José López Portillo, in a major bureaucratic restructuring, combined the two in 1977, creating the Ministry of Agriculture and Water Resources (Secretaría de Agricultura y Recursos Hidráulicos—SARH). More than ninety programs dealing with rural development were placed under the jurisdiction of the new ministry.

The reorganization has necessitated a period of restructuring, plagued by problems of "compartmentalization, centralization, and competitive planning and budgeting" (Bailey and Link 1981:10). López Portillo made decentralization a major goal of his administration, but despite the rhetoric, little power was transferred to the local level. The control of irrigation districts continues to be dictated from Mexico City.

CENTRALIZATION AND LOCAL RESISTANCE

In the process of expanding their power, states centralize processes of regulation, accumulation, and distribution. "Centralization is by definition a process of the shifting of decision-making from a multiplicity of many foci to few" (Adams 1988). Agrarian populations throughout the world, particularly the peasantry, have been profoundly affected by this process, because states, as expressions of power, "enforce appropriation of value by the bourgeoisie" (Wallerstein 1984:5). In many cases, the process has been opposed by local populations, who have struggled to maintain local control.[1]

Since the Revolution, Mexican elites have used the state to control the peasantry while increasing agricultural production. The increase of state power and programs has unevenly penetrated rural Mexico. The major regions of irrigated capitalist agriculture have been the focus of state rural investment and management (Carlos 1981; Whiteford 1986). Yet even within the same region, different communities have been incorporated into the state system differently. In many cases, citizens have demanded state services such as clinics, roads, electricity, and potable water. In other cases, rural communities have been incorporated in government development programs that they neither requested nor wanted.

Since the Revolution the peasantry has been courted for political support and tranquillity. At the same time, government leaders have tried to impose an authoritarian and paternalistic relationship upon the peasantry (Hamilton 1982:143). Institutions such as the National Campesino Federation (Confederación Nacional de Campesinos—CNC), organized to give the peasants a political voice, have become a mechanism of state control over the peasantry (Hardy 1984).[2] The Banco Ejidal, created to give the peasants access to credit, has become an institution of manipulation and domination (Hamilton 1982).

Arturo Warman states that "the principal agent of the exploitation of the peasants is the state, which imposes the general conditions for the distribution of resources, their circulation and valuation, for the dominance of capitalism and its preservation" (Warman 1980:5; all translations are ours unless otherwise noted). Other Mexican authors, such as Guillermo de la Peña, agree, pointing out that it is critical to examine not only the macro-level changes in state policies but also their "implications for regional development" (de la Peña 1981:260).

These issues raise an important question examined in this book: How do the peasants of one region perceive the state, and what do they see as the legitimate role of the state in the region? Our analysis focuses primarily on the ownership of irrigation water, its distribution, and administration. From this we expand our analysis to examine issues of class, ideology, and forms of resistance to state domination.

In an era of macro-level analysis and theorizing, a critical dimension is missed if local-level processes and linkages between levels are ignored. It is at the level of individuals, households, and communities that resistance to capitalism and state centralization is played out. It is at these levels that traditions are kept, local organizations forged, class or ethnic consciousness developed, and quiescence or rebellion chosen. This book is about a people's struggle for control of their livelihood at a time when capitalism and state power have transformed their everyday lives.

A critical dimension of the analysis is the nature of centralization as a concept and how it relates to irrigation. To deal with this phenomenon, we address the following questions:

First, does a sociotechnical activity such as irrigation always have organizational prerequisites, particularly centralized control of activities necessary for large-scale operation over a prolonged period?

Second, what is the nature and location of centralized decision making? Does it imply that particular organizations and individuals nestled at some point within the continuum of local to national structures have absolute control over irrigation activities? And if so, what are the implications of local versus outside control?

Third, does the movement to take control over decision making stem from local pressures to resolve local disputes, or is the process one of co-option and domination by the state as its bureaucratic elites expand control and power?

Fourth, does locally controlled technology, which leads to an increase in the means of production, create greater stratification, or are there local mechanisms that can lead to growth with equity?

Debates about centralization have a long history in the study of rural social transformation and irrigation. One of the key issues in the social science literature on irrigation revolves around whether particular irrigation systems have been subject to centralized control. The theoretical basis for such discussion owes much to Karl Wittfogel's *Oriental Despotism* (1957). Wittfogel proposed that the control of large-scale waterworks is an independent variable in the evolution of centrally controlled states in which the management of irrigation agriculture is the main concern of bureaucracies responsible for planning, construction, maintenance, and operation of such works. Both Julian H. Steward (1949) and Wittfogel hypothesized that such large states or "irrigation civilizations" as those of Mesopotamia, China, Egypt, and Mesoamerica had similar institutional characteristics and passed through similar developmental stages, because of their common need for waterworks in arid and semiarid environments. This became known as the "hydraulic hypothesis," which, subsequently, has been refuted by a number of investigators. Most convincingly, Robert McC. Adams (1966) argued that it was the bureaucracies of existing states that designed and built the large irrigation systems in order to meet the demands for increased agricultural productivity. The bureaucracies administering waterworks were therefore subsequent and administratively subordinate to a larger preexisting state.

The arguments over irrigation and the origin of the state have stopped, but the nature and the degree of state control over irrigation activities continue to be discussed.[3] The central question continues to be whether irrigation as a sociotechnical activity, by its very nature, has certain organizational prerequisites. Is some form of centralized control necessary for capital formation, construction and maintenance of the system, allocation of water, and ensuring smooth operation without disruptive conflicts? Or can these goals be accomplished without centralized control?

The idea of centralization is complex and involves the definition of concepts and the interpretation of data from many sources. In order to make valid comparisons, centralization must be defined and operationalized within specific contexts. For example, the definition must be able to discriminate between the allocation of water rights to irrigators and the resolution of conflict between groups of irrigators. In other words, the consequences of centralization are not uniform but affect different irrigation activities and organizations in ways that must be empirically determined. The interpretation of data is difficult, if not impossible, without knowing the detailed history of the sociopolitical structures surrounding an irrigation system. In this volume we will use the concept of centralization developed by Robert Hunt: centralization occurs where control over the water allocation process is taken over by a politically centralized bureaucracy, effectively replacing local organization or organizations (Hunt 1980; Kelly 1983).[4]

Robert Hunt has pointed out that analytically it is important to separate the influence that individuals at the top of the local hierarchies have on national bureaucracies from their personnel who exercise direct government control and administration. Local leaders or elites often have connections and influence with the government, but this is not the same as formal roles in a local branch or agency of a national organization controlling irrigation water. The previous definitions of political centralization included both the administration of irrigation systems by a central polity and the links between that polity and the local elites, which are, in reality, two separate phenomena (Hunt 1980:176).[5]

Following Hunt (1980), we will call the role of directing an irrigation system, vested in an individual or group, *unification*, and the increasing or decreasing control by the national polity and its representatives, *centralization;* both terms label concepts that are variable, depending on very specific contextual circumstances.

The fact that irrigation systems function in complex societies means that there are structural relationships between groups of irrigators and the sociopolitical institutions of the nation. Irrigation organizations carry on many kinds of activities that neither articulate consistently with other institutions nor seem to remain constant over time. In order to make meaningful statements and then to formulate hypotheses about irrigation's relationship to society, it is necessary to know more about specific techno-environmental relationships, the organization of irrigators, state institutions, local ethnicity, and ideology. J. Stephen Lansing's (1987) study of water temples in Bali provides a rare insight into the articulation of ideology and the material aspects of irrigation activities. Quechua Indians of the southern Peruvian Andes have been shown to adjust terrace size and total irrigated area in response to long- and short-term variation in water availability (Guillet 1987). In the Taita Hills of Kenya, Patrick Fleuret shows that management of irrigation in a small acephalous society is related to distribution of status, population growth, land reform, and modernization (Fleuret 1985).

Construction, maintenance, water allocation, and conflict resolution are distinct activities managed by irrigation organizations. In separate ways, these activities are related to various state agencies: construction and maintenance could be controlled by a ministry of public works and also be related to rural labor organizations; allocation of water rights could be part of legal land tenure and be administered by an agrarian ministry; and certain conflicts could be resolved by the irrigators themselves while others might be referred to the courts or government agencies. The irrigation organizations must arrange and maintain cooperation while at the same time resolving conflicts resulting from competition over scarce resources. These ends are achieved within the context of multiple bureaucracies and political organizations with differing amounts of power and control over irrigation activities.

INTEGRATING STRUCTURES AND THE EXPANSION OF STATE POWER

Despite the importance of irrigation in Latin America, and in Mexico specifically, very few studies have attempted to integrate the analysis of irrigation organization with the study of local and regional social structure. Yet in most regions, social relations and organization are intimately interwoven and are strongly conditioned by the differences in power resulting from unequal control of land and water resources. One of the most important factors that influences control of water resources is the actual or implied control imposed by government policy and administration. Consequently, in regions where water and land resources are a source of power and subject to competition, the study of social organization inherently requires an analysis of the individual roles, households, community, and region, and of the links between local-level processes and national-level policy and programs.

Several key factors influence how national policy is implemented on the local level. First is the nature of the institutional structures, such as local and regional offices of government ministries, that tie regions and localities to centers of national power where policies are formulated and plans for implementation are made. Institutions at all levels are part of hierarchically organized bureaucracies whose function it is to carry out government policies. But the bureaucracies are not neutral intermediaries created by the state; they play a major role in managing systems of accumulation, control, and distribution. Furthermore, as Miriam Wells and Jacob Climo (1984:152) point out, "intermediaries, be they bureaucratic agencies or strategically well situated individuals" can independently generate change.

Local representatives of large national bureaucracies may have tremendous latitude in how they use their derivative power, making them key brokers linking communities to national institutions. We therefore suggest that the nature of the relationships between these local-level bureaucrats and community leaders, as well as between national leaders within their own organizations, is a key variable affecting how policies are implemented. In some cases, local bureaucracies are staffed by local elites; in others, individuals from outside the region with no previous local contacts are appointed. Such factors, in turn, influence the nature of networks both within the bureaucracies and between communities and government institutions.

We further propose that the organization of power and interests at the regional and community level is a critical variable influencing how state policies are interpreted and enacted at the local level. These interests are often controlled by specific individuals and families strategically linked to regional and national government officials (Mares 1980).

How federal programs articulate with local populations affects the nature of political and class organizations of the region and community. Where there is a locally powerful elite, it is possible that government policies and programs may be modified or even co-opted, depending on the nature of local goals and interests.

COLLECTIVE ORGANIZATIONS AND AUTONOMOUS CONTROL

In this book we examine a classic case of local collective organization based on shared economic interests and structured by the coalescence of local values. The irrigation associations of the Tehuacán Valley were organized in the absence of central government activity or support. The water captured by the wells has been instrumental in expanding agricultural production in the region. Like other collective, locally based organizations, these are distinguished by the fact that they are self-managed. In contrast to many other economic "institutions of the state and market, they draw leadership and management skills from their own ranks rather than from a cadre of professionals. The organizing principles of collective organization are voluntary membership, government by agreement and social control by peer pressure" (Bratton n.d.). In general, members of collective organizations resist incorporation into state programs and bureaucratic structures, because incorporation places them at the bottom of a hierarchy.

Although collective organizations are found throughout agrarian regions of the world, political leaders and elites often assume that peasants do not know how to form committees, pool resources, manage funds, or maintain local organizations. Because of these factors, development planning is usually imposed on rural areas without incorporating the population in planning or implementation, often leading to peasant resentment and bitterness, as documented in southwestern Iran (Salmanzadeh and Jones 1981) and in a large project to resettle Sahel drought victims in Kenya (Hogg 1983). In the process of modernizing agriculture and enhancing capitalist development, governments displace many traditional institutions, including collective organizations, and concentrate power in the hands of the central government.

The Tehuacán Valley is significant because the peasants of the region, organized into irrigation associations, have built and continue to operate one of the most highly organized, indigenously generated and maintained systems of irrigation in Latin America. Local hegemony has been maintained despite increased government efforts to regulate the system. The construction during the 1940s and 1950s of the systems that today irrigate 16,564 hectares, most of which are cultivated throughout the year, was accomplished entirely without technical advice, financial investment, or centralized coordination from the federal government. Much of this was built at the same time that the Mexican government was investing up to a fifth of its national budget in the construction of major large-scale irrigation projects in other parts of the Republic.

Because rainfall is undependable and limited in the Tehuacán Valley, most land must be irrigated for either subsistence or cash crop agriculture. The only dependable source of irrigation water is groundwater, which technically belongs to the Mexican government, but can be used by those who invest, construct, and maintain irrigation facilities. The right to use groundwater is independent of either private or ejido rights to land, which in this particular area of Mexico means a complete separation of land and water rights, that is, prior appropriation.

The peasants of the region have developed and refined an ancient technology for tapping groundwater without having to make a large investment in gas-powered pumps. They have banded together to form associations that pool small monthly contributions or labor to construct chain wells called *galerías filtrantes.* A tunnel is excavated in such a manner that it uses gravity to bring water to the surface at a lower elevation, sometimes several kilometers from the starting point.

The associations, known locally as *sociedades explotadoras y distribuidoras de aguas de riego,* manage the system without government interference. They have provided the organizational focus for irrigated agriculture in the valley, linking people who hold land privately or through ejido or communal membership. The associations provide (1) sufficient capital to expedite construction, maintenance, and operation of the irrigation facilities; (2) a formal allocation system; (3) the organizational structure to manage irrigation activities; and (4) the resolution of conflict.

The independent development of a system for the appropriation and distribution of water without government support has had a profound social impact on the valley. One of the key questions of our research was how the expansion of the means of production, in this case, water, influenced local power and social relations. Given the context of the capitalist political economy and the expansion of commercial production in the valley, we hypothesized that there would be increased stratification and proletarianization and the emergence of a regional elite whose power was based on water resources.

The elaborate system was built at significant risk and sacrifice to the agriculturalists. Water was not always discovered when the chain wells were built. In addition, because groundwater was owned by the state, there was always the risk that the federal government would intervene and take control of the privately financed systems. In many other regions of Mexico, the federal government has consolidated its power over most aspects of water resource development and distribution. This has not happened in the Tehuacán Valley; water resources continue to be managed on the community and subcommunity levels. The national government has not disrupted the locally controlled system, because of its apparent equity, the high level of its agrarian productivity, the effectiveness of its management, and its support and approval by the vast majority of the residents of the valley.

Yet, although the Mexican government has not taken de facto control of the water resources of the valley, it has expanded its role in the valley through other institutions, setting the conditions and terms of exchange by controlling prices, allocating credit, and administering ejidos and regional political systems. We examined the degree to which control over water resources influenced access to government, and whether access to government enhanced the probability of gaining control over expanding water resources. The answers to these questions, we felt, would lead us to a better understanding of the broader theoretical concerns about the nature of social power and the dynamics of regional stratification.

Clearly, rural people can and do mobilize in many ways. Yet collective organizations are often fragile because of internal contradictions and hostile reactions from the central government or local elites. Internal problems are generated by the fact that the collective organization is designed for individual gain. The organizations make pooling of resources possible, but they also have the potential to reinforce or create inequalities. Because the associations were never intended to guarantee equity in the distribution of benefits, unequal control of the benefits can reinforce class formation in the community or region.

Collective organizations are developed in the context of state and market systems. When they are regarded as a threat to state or elite interests, they are vulnerable. Hostility to collective organizations can be generated for a number of reasons. In some cases, the local organizations are seen as sources of opposition to the centralization of state power and authority. In other cases, local organizations are perceived as an expansion of power and privilege of specific class interests at the expense of others. In still other cases, they are regarded as potentially subversive to efforts of state power.

In Mexico, as we pointed out earlier, water belongs to the state. Yet water captured by the chain wells in the Tehuacán Valley is owned privately by individuals, who may use it, sell it, or trade it. It is obvious that those who own water rights in Tehuacán are opposed to the intervention of the national government. Yet, equally important, and a major issue in our analysis, is the resistance of those who do not own water rights to state intervention and centralization of water management and distribution.

In our ethnographic description and analysis of the irrigation associations, it will become clear that they are an effective type of local organization. Embedded in the local, regional, and national social structure, their existence and persistence generate an important form of organization and social power. They have the potential to evolve into other types of cooperative organizations for production and exchange. Why they have not been the bases for greater organization is an important issue in this book. Equally important is the ideology of independence from state intervention, even if the intervention could possibly bring about a more equitable distribution of water within the communities of the region.

Our research is based on a regional approach because it was clear to us that the processes that transformed the valley transcended community boundaries. Studying two neighboring *municipios* during the same time period provided us with comparable and contrasting results. Similar in size, ethnic composition, culture, and structural relationship to the state, the two principal *municipios* on which this book is based are San Francisco Altepexi and San Gabriel Chilac—two of the largest communities of the valley.

The major population centers of the two *municipios* are located only seven kilometers apart and share public water from the same spring, La Taza. In 1980 the population of Altepexi was 12,521; the population of Chilac was 15,330. People in both communities are Nahuatl- and Spanish-speaking, although a small percentage of the older people do not speak Spanish. Agriculture is important in Altepexi and provides the principal source of income for more than 50 percent of the population, but small industries such as basket making and clay tile production generate income for more than 35 percent of the economically active population. In contrast, Chilac is a town of agriculturalists and traders, activities that provide income for more than 80 percent of the economically active population.

Both communities are highly stratified, as are all of the other communities in the valley. In both, a small local elite (upper peasantry) controls a significant percentage of the water and land resources, and most of the campesinos (lower peasantry) lack adequate land and water to farm independently. In contrast to many parts of the world, water, not land, is the key means of production.

The ejidos in the two *municipios* were created by very different segments of the local populations. In Altepexi the ejidos control 1,350 hectares, of which 64 constitute irrigated land. The 120 people who own private agricultural lands control 259 hectares of irrigated land, and most are also members of the ejidos. In Chilac, the 79 *ejidatarios* control 267 hectares; the private growers own 351.6 hectares of land. The key difference is that the Chilac ejido does not own any water rights, whereas the ejido in Altepexi does. In Chilac private landowners own the majority of the shares in the irrigation associations. In both communities there are significant numbers of families that depend on agricultural work but own neither land nor water, although the numbers are greater in Chilac. Nevertheless, more than 200 workers from the sierra move to both communities to do seasonal agricultural work every year.

OVERVIEW

In Chapter 2 the reader is introduced to the physical and social environment of the Tehuacán Valley. The chapter is written to show the systemic relationship between social and ecological features describing the physical landscape, the climatic conditions, the villages and the central city, land tenure, agriculture, and irrigation systems. We will also show the delicate balance between demographic pressures and the use of land and water by documenting population dynamics in the valley as a whole and specifically in our two communities. We then show how increased population pressure has affected the availability of groundwater for irrigation.

The long and well-documented prehistory and history of the region presented in Chapter 3 gives us a unique opportunity to place the contemporary analysis in a diachronic perspective. At the same time, the analysis serves more than as just a perspective. The region was the site not only of prehistoric irrigation but also of stratification and elaborate political and economic organization built upon the control of hydraulic resources in ways that were remarkably similar to those used on the arid coast of pre-Hispanic Peru (Farrington and Park 1978; Park 1983).

The Spanish Conquest and colonial occupation had a profound impact on the region, yet the major population centers have remained Indian and have resisted domination by outsiders. In the process, the inhabitants of the valley have fought for land and water and created their own systems for developing and maintaining these resources.

The overview is based on the in-depth archaeological work carried out by the research team of Richard MacNeish and the extensive ethnohistorical investigation by the late Eva Hunt. The early history of the Tehuacán Valley has a significance that transcends the steep slopes of the valley. The valley was the site of the domestication of corn and squash, as well as of early settlements based on irrigated agriculture. The social organization just before the Conquest was stratified, with strong evidence that control over land and water was a crucial variable over five hundred years ago. Some of the irrigation systems and land use patterns of the present carry the imprint from this long and complex history. Contemporary conflicts over access to water when viewed over time will reveal patterns and give clues to relationships between resources, technology, and sociopolitical organization.

Chapter 4 outlines the history of conflict over valley land and water between the Spanish and the Indian communities. The struggle for land and water continued after independence, although it was played out in different forms. Contemporary life in the valley carries the indelible marks of its history. In this chapter we also outline the central features of contemporary Mexican water law and national policy and how these policies were implemented during the agrarian reform. We show, with some detail, the specific outcome of the postrevolutionary distribution of land and water in Altepexi and Chilac.

Chapter 5 examines the expansion of agriculture, the emergence of irrigation associations, the associations' institutional structure, the customary laws by which they are governed, the conflicts, and the resolution of these conflicts. In this section we discuss the way irrigation and the associations have become a critical dimension of community and valley life. Although both *municipios* have irrigation associations, we show how they are structurally similar but have distinct operational and conflict histories.

The subject of Chapter 6 is the growth and stabilization of a regional elite with disproportionately greater control over water resources than other members of our two communities have. We ask if this is a result of a decentralized irrigation system, or if it is due to broader forces of the political economy. We then address the significance of this process for the people of the valley. In this context we discuss the integration of roles between those who control water through the *galería* system and their control of other resources and institutions in the valley.

With economic growth based on expanded water resources, new tensions and conflicts have emerged, but not to the extent we had expected. The reasons for this are examined in Chapter 7. Central processes described in this chapter are migration, trade, and the nature of class and ethnic conflict. How these factors relate to the expanded water system and to stratification is discussed.

In Chapter 8 we compare some of the processes examined in the two *municipios* and discuss the implications for our propositions and general model. We then make some comparisons with the impact of irrigation on local organizations, national bureaucracies, and the implementation of policy in other parts of the world. In the process, we reexamine the concepts of centralization, appropriate technology, and development. Of critical importance is a redefinition of what is called locally controlled development and the factors that influence its social impact.

2.
THE TEHUACÁN VALLEY

SURROUNDED BY towering mountain ranges starkly outlined on the horizon, the Tehuacán Valley stretches like a verdant carpet crisscrossed with brown stitching. The steep escarpments glow with color as the sun rises to burn down on the valley, later to exit as the winds swell and twist the dust on the fields and roads into swinging columns. Hundreds of kilometers of canals, often lined with elegant stands of *caña brava* (a kind of bamboo used in basket making), direct the flow of water from the springs, which for more than two thousand years have made agriculture and settlement possible here. The calcified ruins of old canals stand as testimonials to the years of labor that tapped and guided the water generations ago. Invisible to the eye, except for the string of open holes in the ground, is a vast network of hand-dug tunnels, tapping veins of water and channeling them to the surface at lower elevations. A relatively recent modification, the *galerías filtrantes* are a major new source of water and have contributed to transforming the social organization, as well as the physical features, of the valley.

At first glance, the dusty streets of the towns hide the signs of change. But the grinding gears and rumbling diesel engines of the omnipresent trucks that link the towns with other parts of Mexico are artifacts of a new way of life that has encompassed the region. Between the towns, rich fields of corn, beans, tomatoes, and sugarcane grow, dependent on irrigation for survival. Equally dependent on the precious water and the land are the thousands of families that have sacrificed to build and maintain the elaborate irrigation system without the help of the central government.

ENVIRONMENTAL AND DEMOGRAPHIC CONDITIONS OF THE REGION

The Tehuacán Valley, as shown in Figure 1, is located in South Central Mexico, approximately 180 kilometers southeast of Mexico City, in the states of Puebla and Oaxaca. The general orientation of the valley is north-northwest to south-southeast. At the northwestern end, the valley is a southern extension of the Puebla Basin, a part of the Central Mesa, and here the valley is broad and high with low borders. The width at this northern end is about 15 kilometers and the altitude is 2,200 meters. The terrain drops gently but steadily from the north to the southeast; the total length of the valley is approximately 100 kilometers. At the southern end, the altitude drops to 800 meters. The average gradient of the entire length is 14.6 meters per kilometer.

The Tehuacán Valley occupies an area a little over two thousand square kilometers, or about 8 percent of the territory of the state of Puebla. According to a 1981 study of land use, 41,103 hectares were under cultivation, of which 16,564 (40.2 percent) were irrigated, 24,564 were cultivated without irrigation, 49,348 were used for grazing land, 2,191 were forest, 63 were *ociosas* (empty), and 2,832 were unproductive (SARH 1982:28).[1]

The city of Tehuacán occupies a position in the middle of the valley at an altitude of 1,700 meters. From Tehuacán to Teotitlán in the south, the distance in a direct line is 48 kilometers. The gradient of the southern part of the valley is about 19 meters per kilometer. This is misleading, however, as the first 300 meters consist of the descent from the Tehuacán terrace to the Llano of La Taza, and the next 133 meters come in the descent over its edge in two steps, the hill of San Marcos and of Altepexi. These are separated by long and slowly descending gradients of 8 to 10 meters per kilometer. From Altepexi to Teotitlán, the descent is even more gradual (Byers 1967*b*:46).

To the north and east, the valley is bordered by the Sierra Zongólica, which forms part of the northern end of the Sierra Madre de Oaxaca. To the southeast is the Sierra Mazateca, which reaches altitudes of 2,500 to 3,000 meters, with individual peaks rising to 3,500 meters. The southern and western sides of the valley are formed by the Sierra de Zapotitlán and other ranges that border on the Mesa del Sur, commonly known as the Mixteca Alta or the Sierra de Mixteca. These mountains have a maximum elevation of 2,500 meters.

The slopes of the eastern mountains, the Sierra Zongólica, present a rugged appearance with limestone cliffs and depressions made by numerous small watercourses that have eroded the surface. Because these eastern mountain ranges rise in the path of moisture-bearing

Figure 1. The Tehuacán Valley, Mexico

winds from the Caribbean, they receive more precipitation than the Sierra de Zapotitlán to the west. The large number of streams that flow down the mountains to the east are a consequence of the differences in rainfall. Occasionally these streams have cut deep into the mountain slopes to form relatively wide and long valleys, several kilometers long and nearly half a kilometer wide. The Río Salado, the principal river in the valley, flows southwestward from southern Puebla to northern Oaxaca. The Río Zapotitlán traverses the highlands to the west of the valley and joins the Río Salado in the southern part of the valley (Figure 2). These rivers carry surface flow during the rainy season, but during the rest of the year the flow is subterranean. The lesser tributaries carry water only after a large rainfall. At the southern end of the valley, the Río Salado joins the Río Grande, which flows north from Oaxaca. Together, they form the Río Santo Domingo, which cuts a narrow valley, with steep cliffs and embankments, through the Sierra Madre de Oaxaca and flows eastward, where it becomes the Río Papaloapan. This large river provides the major drainage for the eastern side of the mountains, which forms the border of the Papaloapan Basin on the coastal escarpment in the state of Veracruz.

The valley was formed during the early Quaternary era when a lacustrine basin and lake, 1,700 meters above sea level, was drained. The basin began to drain when water started to flow east after the formation of the Río Santo Domingo, which made a connection to the Papaloapan Basin. These events led to the disappearance of the basin and the streams that led from it, resulting in higher levels of erosion and a general increase in drainage of the valley. In essence, the lake, originally a closed system in equilibrium, was drained and transformed into an area of intense erosion. Today the drainage and erosion continue as the reserves of water in the surrounding mountains have continued to decrease, causing the hydrostatic groundwater base level to sink lower as time passes. The net result of this continuing geological process is that the valley is becoming drier and more desertlike (Brunet 1967:74–75).

The regional topography has provided both the barriers and the opportunities for populations that have occupied the Tehuacán Valley from prehistoric times to the present. Tucked between towering mountain chains, the rich valley has been a center of growth and trade as well as an avenue linking important cultural and production centers. It connects Oaxaca with the heavily populated Central Plateau. The interaction between these two regions dates back thousands of years. The regional capital, the city of Tehuacán, is located 254 kilometers from Mexico City and 311 kilometers from the city

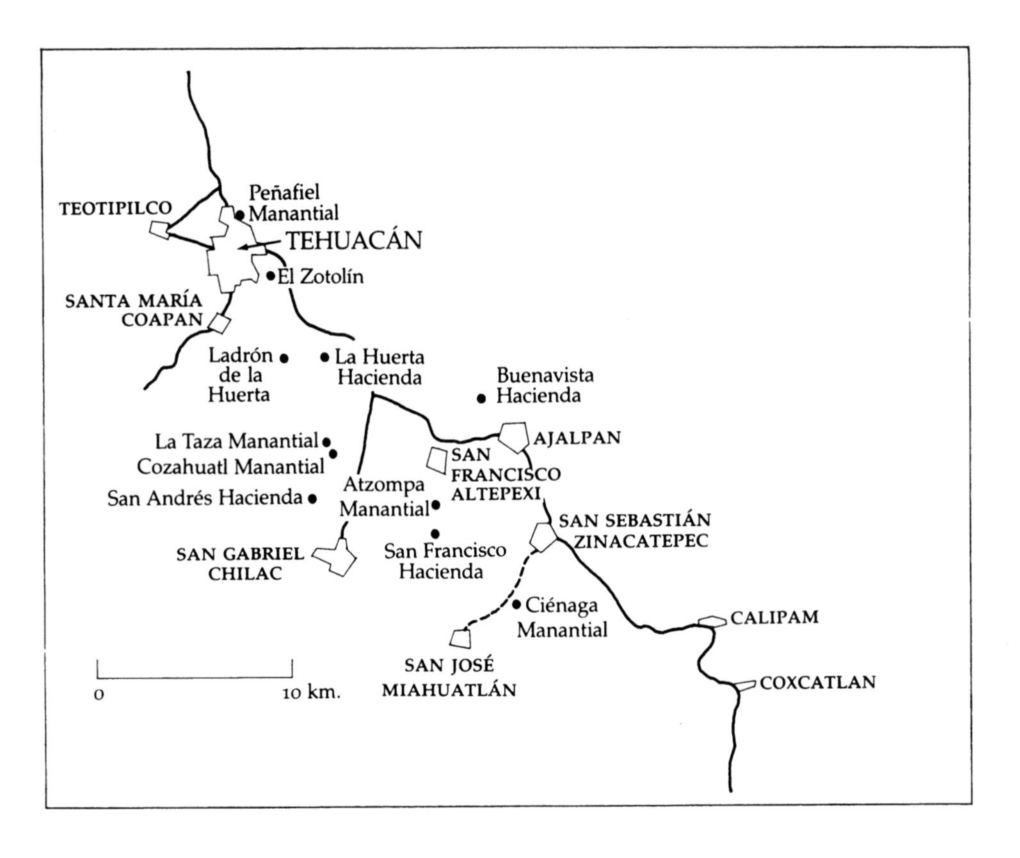

Figure 2. The Southern Tehuacán Valley

of Oaxaca. The Mexico City–Oaxaca railroad, constructed in 1892, passes through the valley, facilitating the transportation of produce into and out of the valley. The valley is also linked to the Veracruz lowland and southeastern Mexico by a steep twisting highway. The city of Orizaba, located on the road to Veracruz, is only 67 kilometers from Tehuacán, and the port city of Veracruz lies 331 kilometers away. The lowland area is an important market for products from the valley. Many Tehuacán families have moved to Veracruz and cities in the Southeast. The highland communities of the sierra have historically been linked by trade with the people in the Tehuacán Valley. Occupying very different ecological zones, the sierra and the valley communities have a long tradition of trade.

The urban center of the valley is the city of Tehuacán, a thriving commercial center. Between 1960 and 1970, its population grew 51.3 percent, reaching 68,332; today it is approximately 130,000. This city is known in Mexico for its bottling companies, such as Peñafiel, El Riego, Balseca, and Aguas de Tehuacán, which have converted the local springwater into a major commercial enterprise. Although its health resorts continue to attract health-conscious tourists to the city to bathe in the springs, new concepts of health along with the development of modern Mexican resorts have left Tehuacán as a tourist backwater. The elegant old resorts are a reminder of times gone by, when Tehuacán was a popular health retreat.

For the people of the valley, the city is the major commercial, administrative, and service center in the region. The major banks, educational facilities (including the Instituto Nacional de Indigenismo), and hospitals of the region are located in the city. The Tehuacán markets and stores thrive on the commerce from the communities in the valley. They are the major distributors of industrial goods ranging from radios to beds. The expanding market for these products is controlled almost exclusively by these merchants. Although the indigenous peoples of the valley control most of the sales of agricultural produce, traders based in Tehuacán are the major buyers of coffee in the sierra.

Political and economic power in the region is no longer controlled by the old families that once dominated the valley. Today much of the power rests in the hands of Spanish, Lebanese, and Mexican families that have moved to the city since 1940. These families have developed a variety of businesses, ranging from clothing factories employing large numbers of women from the valley to major commercial enterprises.

The business that generates the most revenue and employs the largest number of people is the Beneficiadora Avícola Tehuacán y El Calvario, a highly sophisticated agribusiness complex. This family-owned enterprise has farms strategically located throughout the valley, where both local and migrant laborers raise chickens, pigs, and cattle and tend the vineyards. Controlling more than three-thousand hectares in the region, the company has built a new feed-processing plant in addition to its modern marketing and processing facilities for eggs and chickens. A horizontally integrated company, the operation is one of the largest egg producers in all Mexico.

Although the valley is blessed with a frost-free climate, agricultural production is nevertheless constrained by the limited, unpredictable rainfall. The climate of the Tehuacán Valley is semiarid, with substantial differences in seasonal precipitation and moderate annual variation in temperatures. Although the temperature records are not complete, the mean monthly temperature for 1956 to 1965 was 20° C, and the mean minimum and mean maximum temperatures were 17° and 25° C, respectively. Average monthly temperatures in different areas of the valley do not vary more than 3° C, but extreme day and night differences do occur, and a 30° variation is not unusual. The lowest monthly mean temperatures occur in December and January, and April and May are the warmest months. Winter storms in the United States often send waves of cold air, called *nortes*. *Nortes* rarely bring rain to the Tehuacán Valley, but they do bring cool winter weather (Byers 1967*a*:50).

Most of the rainfall occurs during one season, normally beginning in early May, peaking first in June, diminishing toward the end of July, and peaking again in September (Table 1). This latter period, the *veranillo*, or little summer, is when the rains from tropical storms provide additional moisture for crops that are otherwise not irrigated. Lack of these rains would spell certain failure to unirrigated crops. Severe downpours, however, rapidly erode the upper layer of the topsoil, resulting in irreparable damage to the crop. If planting has just taken place, the seeds are washed away, and more mature plants lose their root support and topple over. From early November through April, there is normally little or no rain (Byers 1967*a*:57; Thomas 1968:5).

The topography of the valley has a strong influence on precipitation. The Sierra Madre de Oaxaca shields much of the valley from storms approaching from the east. This shield is more effective in the northwestern part of the valley than in the southeast. The annual average rainfall from 1926 to 1955, over a large area of the northwestern valley, was only 500 millimeters, the highest aridity being centered around Tehuacán. During the two peaks of the rainy season, the average weekly rainfall in Altepexi is only 20 millimeters per week (Byers 1967*a*:54). Precipitation of 250 to 450 millimeters per year, common in the center of the valley, is characteristic of desert climates. Droughts are common, and were documented for three of the summers during the ten-year period from 1950 to 1960 (Comisión del Papaloapan 1965:14). A similar pattern of dry years has been recorded for the two subsequent decades. In general, areas that are situated near the foothills and mountains in the eastern part of the valley receive more rain than those in the center (Figure 3).

In this arid climate, where evapotranspiration exceeds precipitation throughout the entire year and the rainfall is limited and unpredict-

Table 1. ***Precipitation in the Tehuacán Valley, 1955–1975***
(Altepexi Observation Station)

	Jan	*Feb*	*Mar*	*Apr*	*May*	*Jun*	*Jul*	*Aug*	*Sep*	*Oct*	*Nov*	*Dec*	*Annual*
1955	0.1	0.0	0.0	7.2	3.2	47.8	168.5	60.0	255.9	36.0	9.0	1.0	588.7
1956	0.0	2.0	0.0	2.5	63.8	143.4	44.8	8.7	39.0	0.1	6.4	0.2	310.9
1957	0.0	23.0	0.1	0.0	107.9	42.5	35.0	45.9	35.2	0.8	0.0	0.0	290.4
1958	51.6	0.2	0.0	9.2	94.2	47.9	72.9	28.3	231.6	66.5	5.7	6.0	614.1
1959	2.0	0.0	0.0	29.0	30.2	207.5	69.0	52.5	57.8	95.0	1.5	0.0	544.5
1960	0.0	0.0	0.0	0.0	3.0	44.5	38.0	66.0	173.5	20.0	0.0	0.0	345.0
1961	0.0	0.0	4.0	0.0	5.0	85.5	67.0	12.0	229.5	0.0	26.5	8.0	437.5
1962	0.0	0.0	0.0	14.0	0.0	82.0	22.0	40.0	106.0	23.0	0.0	0.0	287.0
1963	0.0	0.0	0.0	5.0	17.0	71.0	69.0	41.0	31.0	6.0	7.0	0.0	247.0
1964	4.0	0.0	0.0	13.0	146.0	70.0	43.0	46.0	57.0	0.0	20.0	0.0	399.0
1965	6.0	7.0	25.0	14.0	9.0	86.0	22.0	11.0	37.0	16.0	0.0	10.0	243.0
1966	0.0	11.0	23.0	10.0	103.0	69.0	33.0	45.0	77.0	109.0	0.0	0.0	480.0
1967	6.2	0.0	7.0	10.0	20.0	48.5	8.0	49.5	121.0	77.0	0.0	0.0	347.2
1968	13.0	0.0	0.0	56.0	106.0	109.0	28.5	29.3	13.0	19.0	4.5	2.0	380.3
1969	41.0	6.5	27.5	11.2	17.0	18.0	40.5	253.8	41.0	6.5	0.0	0.0	463.0
1970	0.0	0.0	0.0	0.0	63.0	43.5	59.5	112.5	90.5	0.0	0.0	0.0	369.0
1971	0.0	0.0	0.0	3.5	2.5	27.5	43.5	27.0	59.0	64.5	6.0	0.0	233.5
1972	0.5	0.0	0.0	21.0	49.5	65.0	15.0	65.5	40.0	0.0	0.5	0.0	257.0
1973	0.0	1.0	0.0	6.5	38.0	174.0	103.0	18.5	93.5	0.0	8.0	0.0	442.5
1974	0.0	0.0	0.0	0.0	0.0	—	15.0	5.5	80.5	0.0	0.0	0.0	—
1975	0.0	0.0	0.0	0.0	160.0	36.5	65.1	104.0	95.5	53.5	0.0	0.0	514.6
Average	5.9	2.4	4.1	10.1	49.4	72.3	50.6	53.4	93.5	28.2	4.5	1.2	389.7

Source: Secretaría de Agricultura y Recursos Hidráulicos, Comisión del Papaloapan 1975*a*.

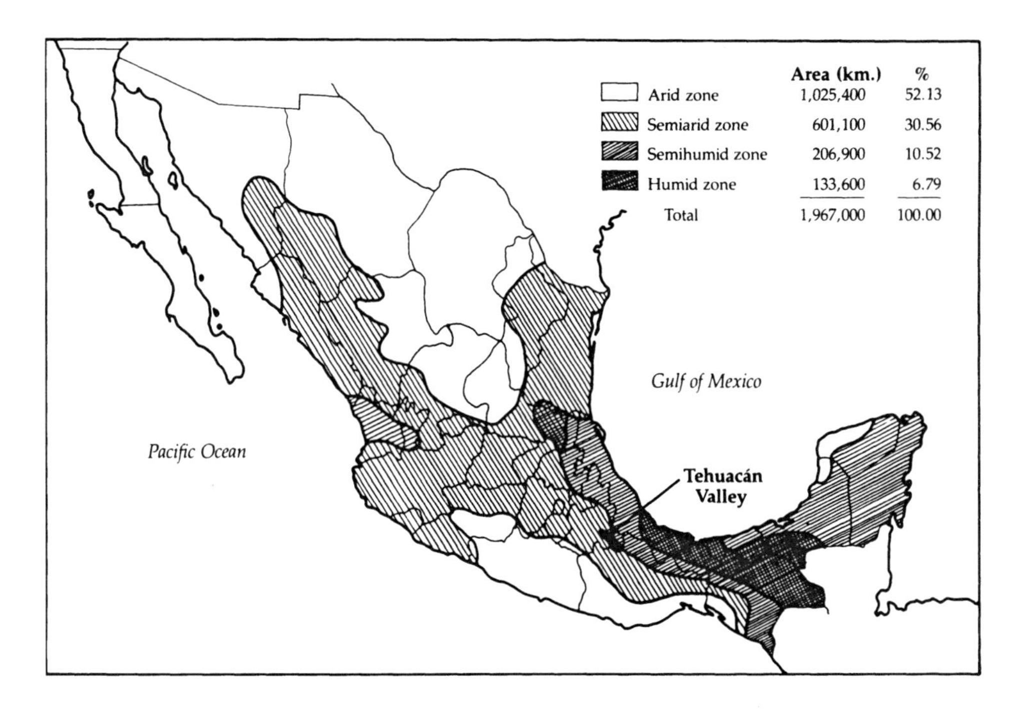

Figure 3. Arid, Humid, and Intermediate Precipitation Zones (after Orive Alba 1970)

able, agriculture has to be supported by irrigation. Low-yield single crops can, at times, be supported by rainfall, but the risks that must be taken by the agriculturalists are considerable.

Agricultural yields in the Tehuacán Valley have increased significantly with the use of fertilizer. The soil is derived from the limestone rocks that border the valley and that are largely responsible for the high concentrations of calcium and carbonate compounds. The alluvial soils were washed into the valley and deposited during its early geological stages (Stevens 1964:278). In addition, the seasonal floods of the Zapotitlán and Salado rivers deposit new layers of alluvial soil, making the land along the river valleys the best for agriculture. In some areas, these lands have been cultivated for the last two thousand years (Flannery 1967:132).

Irrigation systems actively influence the fragile ecology and, in many cases, threaten the productivity of the soil. In the Tehuacán Valley, two significant problems are a major source of concern for the agriculturalist: salinization and the formation of hardpan just below the surface. Salinization is created by the gradual concentration of excess calcium salts and carbonates deposited in the soil by water used for irrigation. The salt buildup, if it reaches high levels, can be toxic to plants and can lead to abandonment of fields. This process has destroyed irrigated fields throughout the world. On lands along the rivers, seasonal floods wash away the salt deposits in the soil, but in other parts of the valley, increasing salinization is a growing problem.

The concentration of calcium, if high enough, can form a hard, rocklike layer beneath the surface, called caliche. The buildup of caliche makes proper drainage of the soil nearly impossible. Calcification occurs in climates like that of Tehuacán, where evapotranspiration exceeds precipitation, resulting in little or no leaching of metallic ions. In the case of the Tehuacán Valley, calcium carbonate in solution is carried in the irrigation water as well as by capillary action during periods of little surface moisture. The calcification contributes to the salinization problem. Despite these problems, careful management of soil and water resources has enabled the people to maintain agriculture.

This ethnography is based on research carried out in the southern portion of the Tehuacán Valley, which includes the contiguous *municipios* of San Gabriel Chilac, San Francisco Altepexi, San Juan Ajalpan, San José Miahuatlán, and San Sebastián Zinacatepec.[2] Although there are important differences between the *municipios* and communities within the *municipios*, the region shares a history and culture.[3] Equally important for our study, *municipios* in this region are linked through irrigation systems that cross *municipio* borders and that are run by management organizations that often include members of different communities.

Many of the agricultural communities in the southern part of the valley have histories going back to loosely organized chiefdoms and independent small villages or city-states prior to the conquest of the Aztecs at the beginning of the sixteenth century. These pre-Hispanic peoples were spread over large geographical areas and spoke a variety of languages.

Ethnic identity, however, was and continues to be based more on the local community than on region, language, or any form of national culture. The pre-Hispanic population was clearly composed of Indians, but the Conquest initiated five hundred years of social interaction that resulted in increased phenotypical uniformity. The current distinction between Indian and mestizo is somewhat problematical. The "Indians" continue to identify with specific villages and to speak recognizable dialects of Nahuatl that are specific to places separated by only a few kilometers; mestizos, who speak only Spanish, very often have parents, friends, and relatives who are ethnically Indian. The post-Conquest history of the valley has been one of conflict between the Indian communities and the Spanish over land, water, and trade. Competition for resources is still a dominant dimension of life in the valley, where a mixed population currently controls most of the land, water, and trade.

Agricultural land in the scattered valley communities is not wasted. Generally, houses are tightly clustered in nuclear settlements. The walls of the houses line the dusty streets, disguising major differences in wealth between families. The most common building materials are adobe and cement blocks, and the backs of the houses often open out onto a cleared area where oxen, burros, and other animals are kept. Overhangs cover courtyards that are used for storing and drying corn, garlic, or other crops.

Electricity and running water are relatively new to the larger communities. Many homes still have neither electricity nor running water, and public fountains or faucets still provide water. The main public buildings that play an important role in community life are the churches, schools, and municipal centers. Each town has its own market day, although the major markets in the valley are in Ajalpan and Tehuacán. During the day, the people work in the fields as well as in domestic activities. Early morning and late afternoon are times of great activity in the streets, as trucks, oxen, and herds of goats returning from the fields interrupt the play of children, who scuttle into open doorways as the shadows of the intruders darken the streets.

None of the major communities of the region (Altepexi, Ajalpan, Chilac, or Zinacatepec) is strictly agricultural. Beautifully embroidered blouses are made by women throughout the region and are sold in elegant shops in urban centers or exported abroad. Tortilla mills, blacksmith shops, street corner grocery stores, tailor shops, and even an occasional furniture store open out onto the main streets. Baskets made of *carrizo*, a tall reed that grows along irrigation canals in Ajalpan and Altepexi, are sold throughout Mexico and are exported to the United States as laundry hampers and Easter baskets. Tiles and bricks are made in a small factory in Ajalpan, and woodworking shops produce a number of different products (Henao 1980:64). Despite this commerce, however, agriculture is the lifeblood of the people of the Tehuacán Valley.

The complex land tenure pattern of the valley is overlaid by a different pattern, a network of irrigation canals that dot the surface. Ejido and communal lands make up 86 percent of the arable land in the valley (SARH 1982:42).[4] But in the southern valley, almost 40 percent of the irrigated land (2,314 hectares) is privately owned. The irrigated land is significant because this is the land that is the most productive and the most heavily cultivated. Because of the old tradition of separating water rights from land rights, ownership of irrigated land does not mean ownership of water; it only means that the land is located in a region that can be irrigated and is probably being irrigated. As we shall see, there are many ways in which people with land can gain access to water for irrigation if they do not own water rights.

Not all of the campesinos in the valley have access to land. One study found that 10,219 day laborers did not have access to land and that there were 10,171 campesinos, including *ejidatarios*, with land (SARH 1982:43).[5] Private landholdings are increasingly concentrated in the hands of a limited number of households. For example, in Ajalpan in 1970, 7.6 percent of the heads of households had five or more hectares of land. They controlled 76.9 percent of the private land cultivated, and 60.2 percent of the privately owned irrigated land (Veerman n.d.:74).

IRRIGATION

Irrigation water in the valley has historically come from a series of springs (*manantiales*) that flow out onto the valley floor. These springs have been a major source of conflict between communities and the haciendas. Today, the most important springs are La Taza (Texcali), which is divided equally between the communities of San Gabriel Chilac and Altepexi; Atzompa, which is owned by Altepexi; Cozahuatl, owned by an association in San José Miahuatlán; and Tochotl, controlled by San Sebastián Zinacatepec (Figure 2). In addition, other large springs are located in the city of Tehuacán, or in the immediate outskirts: El Riego, Peñafiel, San José, San Jorge, San Miguel, San Lorenzo, Cruz Roja, Santa Ana, and Los Alamos.

Community ownership of water has a long history in the valley. For example, ownership of water in La Taza dates back to 1543, when the community of San Gabriel Chilac was founded. The community grew in four stages as different bands moved into the area. The founders of Chilac allocated water to each of the groups. As the community grew, each group became a barrio and each administered the distribution and management of water. The amount of water that was allocated to a landowner depended on the amount of land owned. Later, when community fountains were built, each barrio took the responsibility for collecting maintenance fees from every family in the barrio (Gil Huerta 1972: 154–161).

Irrigation has been practiced in Spain since the occupation of the Iberian Peninsula by the Romans. The Moorish conquest of Spain resulted in the introduction of new practices and regulations, which were later incorporated into Spanish water law (Butzer et al. 1985). In the New World the Spanish imposed their water laws on indigenous populations. According to Spanish law, water was allocated according to the amount of land an individual irrigated. People were responsible for maintaining the system in direct proportion to the percentage of water they used (Dunbar Ortiz 1980: 55). Water rights were allocated to those who first began to use the water. Water in excess of the water used to irrigate could be sold to the highest bidder. In other situations, customary law required that senior water owners had to allocate water to other members of the community in times of hardship (Glick 1972). Since regulation rested in the hands of those who used the system, irrigation management was kept at the local level and out of the bureaucratic structure of the state. Today, the administration of water is coordinated by the Junta de Aguas Federales del Manantial La Taza, which distributes the water to ejidos in Altepexi, Ajalpan, and Chilac.

A second, but irregular, source of water in the valley comes seasonally from the *barrancas* (streams that flow only when it rains). Based on a survey of resource management systems in Middle America, Gene Wilken concludes that the Tehuacán Valley has "undoubtedly the most extensive and complex runoff management system in Middle America" (1979: 18). The backbone of this sytem is the Río Zapotitlán, which is joined by the Barranca de San Antonio just above Chilac. During the rainy season, between May and September, heavy rains in the sierra fill the *barranca* with *aguas broncas* (wild water). The waters are partially diverted by the agriculturalists from San Gabriel Chilac and San José Miahuatlán, who have built a series of diversion dams that stretch across the *barranca*. The dams force water into the network of canals that carry the water to the fields. This water is particularly important because it contains rich topsoil, which is deposited on the fields, making them the most productive in the valley. Water from the first storm is not used for irrigation, to allow the salt and debris that have accumulated in the streambed during the year to wash away.

For the villagers who have land along the *barranca*, the five or six major storms that fill it with water each year represent a major opportunity for obtaining water, but also entail a risk. When the storms come—often at night—the agriculturalists must run to the fields with flashlights and lanterns to open the gates (*compuertas*) and clear away the earth, rocks, and branches that may block the flow of water to the fields. San Gabriel Chilac has twelve major *zanjas,* or canals, that are fed by the *barranca.* The current of the stream is very strong and the water deep, making the work on the banks dangerous, especially at night. Men have been swept away by the current and drowned in the turbulent water. Despite concrete-reinforced borders, especially on curves, there is a serious erosion problem along the *barrancas.*

The work has to be coordinated because it involves not only opening the gates, but controlling a series of other canals. The water rushes down the main channels, which stretch as far as five kilometers. A maze of secondary and tertiary canals then guides the water to the fields. The fields are divided into *pantles* (divisions of land with earth borders). *Pantles,* the basic units of irrigation, range in size from less than an eighth of a hectare to a quarter of a hectare, and they take many shapes, depending on topographic and ownership patterns. Fields that are watered with *aguas broncas* have to have thick borders to sustain the impact of the flood-water. Some of them may be up to four feet thick at the base and more than three feet high. When the *pantles* are filled, the gate is closed, allowing the water to soak into the soil and leave a new layer of soil. The process has to be coordinated because too much water can wash out the gates and the fields, destroying years of work.

The basic units of organization are the *sociedades de zanjas* (canal associations), functioning as cooperative labor organizations. All the people who own land irrigated by the canals belong to the associations and contribute labor or funds to maintain the canals. The different associations are coordinated so that the canals farthest upstream are allowed to get water first and those downstream must wait until there is enough water to guarantee the upstream irrigation. Although there is usually plenty of water, some storms do not provide enough for the people of San José Miahuatlán. The allocation system is important for preserving harmony in the region. The Tehuacán system is unique, "since most runoff irrigation systems lack sufficient control over water supplies and timing to have allocative elements" (Wilken 1979:19–20).

Because of irrigation schedules and the unpredictability of the rains, there are times when farmers do not want to use the water from the *barranca.* The best time to use it is when the fields are fallow and the major runoff is coming from the upstream region of Zapotitlán de Salinas, where the soil is of better quality than it is in other areas.

When the crops are growing, too much silt can damage them or the timing of the flow may not correspond to the needs of the crop schedule. Members can choose whether or not to irrigate, depending on the state of the crops.

The third and most important source of water today in the Tehuacán Valley comes from the *galerías filtrantes*. The *galerías* are a network of subterranean tunnels that are linked to the surface by a series of vertical shafts or wells. They are constructed so that they use gravity to drain groundwater from areas up the valley to a point at a lower elevation where the tunnel emerges at the surface.[6] The Tehuacán Valley is an ideal location for operating the *galerías* because of the gentle slope of the valley and the relative proximity of abundant, high-quality groundwater. *Galerías* are constructed in a southerly direction at a slope that is slightly less than that of the valley floor; thus the tunnel intersects the ground surface if it extends far enough.[7] The length of a single system is highly variable and may extend up to 12 kilometers. Richard Woodbury and James Neely (1972:143) estimate that the total underground length of all the *galería* systems in the Tehuacán Valley is about 230 kilometers. The average length of one of the 105 systems they observed in the valley is 2.2 kilometers. The engineers at the local Papaloapan Commission office in Tehuacán claim that the total length is well in excess of 600 kilometers, but this figure may include systems that are no longer in use. Figure 4 shows the approximate location of the major *galería* systems in the valley.

A 1981 study of the geology and hydrology of the valley reached significant conclusions about the nature and location of groundwater. Two distinct geological strata contain aquifers with water resources and, in some areas, show promise of potential reserves for the expansion of irrigation agriculture. Both strata are infiltrated by rainfall on the porous rock formations of the Colorada and Las Cruces sierras, which lie directly to the northeast of the city of Tehuacán and form part of the Sierra Zongólica. The Sierra Zapotitlán to the southwest, however, is composed of less porous rock and receives less rainfall, contributing relatively little to groundwater infiltration (Geo-Re 1981).

The two strata containing aquifers are located at depths of 30 to 50 meters and 114 to 195 meters, with dense, relatively impermeable layers separating the two. The shallowest stratum is the principal source of water and is tapped by *galerías* and wells. Its composition is primarily an alternation of coarse and fine gravel in conjunction with nonconsolidated sediments of calcium and chalk. This stratum, part of the Formación de Tehuacán and the Cerro de la Mesa, is continually recharged and therefore contains free aquifers. The second,

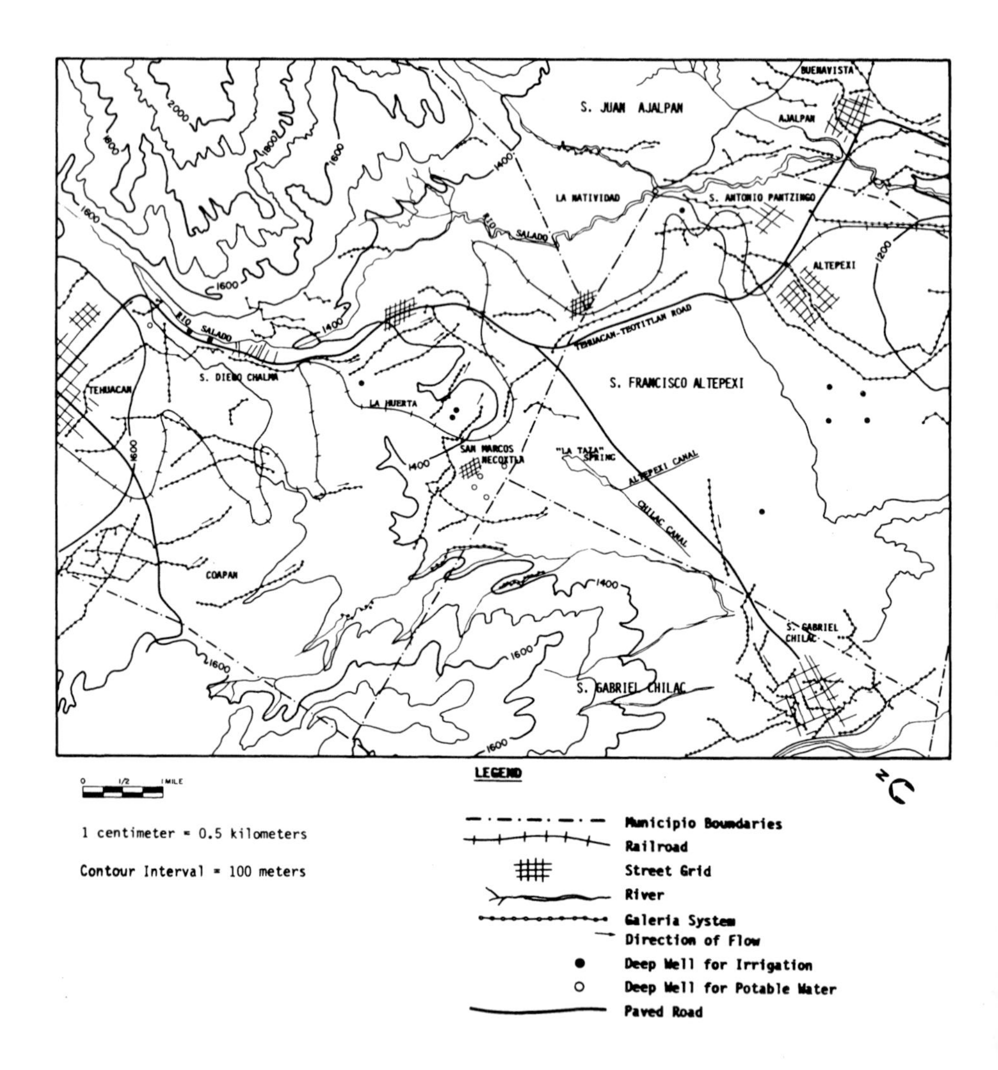

Figure 4. Distribution of *Galerías Filtrantes* in the Tehuacán Valley (after Secretaría de Agricultura y Recursos Hidráulicos 1981)

deeper stratum is composed of fine gravel, conglomerates, sand, and chalk deposited on the lake bottom during the Quaternary. The aquifers are semiconfined and confined with some lateral variation, but essentially with negligible recharge, providing a nonrenewable and finite resource (Geo-Re 1981).

The natural springs, or *manantiales*, are not simply the surface intersection of aquifers in the shallow stratum. Instead, they result when a portion of the water percolating from the mountains encounters impermeable strata in the Formación de Tehuacán and the Formación Cerro de la Mesa. Fractures in these strata act as conduits for water under hydrostatic pressure from the mountains, thus forcing flows to the surface. Extensive mineralization occurs when this water passes through Tertiary and Quaternary volcanic rocks (Geo-Re 1981).

The amount of water in the two strata is localized in specific parts of the valley. The shallow stratum has the most abundant aquifers southeast of Tehuacán close to the eastern sierra and extending from San Pablo Tepetzingol to San Sebastián Zinacatepec. The deep stratum is more extensive to the north of Tehuacán to the *municipio* of Tepanco de López and south to Ajalpan, Altepexi, and Chilac. Further studies have been recommended for the region near San Pedro Tetitlán, Zinacatepec, Calipan, and Coxcatlán (Geo-Re 1981).

Geo-Re made a number of recommendations to government planners in 1981. The deep stratum should continue to be explored by drilling wells, known locally as *pozos profundos*, using modern technology. About thirty wells have been drilled with mixed results. Although a few have good flow rates, most have had disappointing yields. The report goes on to say that well drilling should be carefully planned and reliable lithological and electrical measurements should be made. In order to eliminate problems caused by mud formed from nonconsolidated sediments sealing the aquifers, equipment using percussion or air rotation would produce the best results. Furthermore, as wells are drilled through the shallow aquifers, the first forty-five meters should be sealed to prevent interference with flow in the *galerías*. Extensive work should be done to rehabilitate existing *galerías* so as to continue using the renewable flow in the shallow aquifers.

A number of reservations were expressed about making new *galerías*. Reasons included the high cost of construction, overexploitation of aquifers, the constant deposition of travertine inhibiting a constant reliable flow, and the constant lack of maintenance on the part of local users. If new *galerías* were constructed, they should be in the southeastern part of the valley, close to the Sierra Colorada. Other recommendations urged the immediate determination of aquifer capacity in order to resolve the politics of whose lands would be irrigated and which industries would benefit. Current water use should be accurately determined in relation to socioeconomic organization and history (Geo-Re 1981).

Because the depth of the groundwater is quite variable, the vertical shafts of the *galerías* range from three to seventy meters. Construction has traditionally been done using manual labor, and *galería* systems have taken years to complete. In the Tehuacán Valley new systems are constantly being constructed and those in operation are being maintained. According to George B. Cressey (1958), the essential idea is to excavate a gently sloping tunnel that extends upslope until the water table is tapped and water emerges at the downslope end. Often, such tunnels skirt the periphery of an alluvial fan. To give access to the tunnel, vertical shafts are dug at intervals of thirty to one hundred meters. Where the tunnel dips below the water table, it serves as an infiltration gallery and may have several branches to increase the inflow. The lower and longer part is the conveyor channel. Figure 5 shows a *galería* system typical of the large number that have been constructed in the valley.[8]

In order for *galería* systems to function, the grade or slope of the terrain must not exceed 3–5 percent. Higher slopes result in excessive water speed, which is difficult to control and will erode and damage the canals. According to the U.S. Soil Conservation Service, the ideal grade is 0.8–2.7 percent (Robins and Rhoades 1958). The maximum grade of the Tehuacán Valley is well below that, and the terrain between Tehuacán and Altepexi is within the ideal range.

In 1978, there were 129 *galerías* that irrigated a total of 16,539 hectares, producing more than 400 million cubic meters of water annually. More than 90 percent of the water used for irrigation in the valley now comes from *galerías* (SARH 1982:34). Yet as recently as 1958, *galerías* produced only a small fraction of the water in the valley. The cause and the social consequences of this expansion are central subjects in this book. The antiquity and dissemination of the *galería* system in Mexico and, more specifically, in the Tehuacán region is still a subject of debate.

It has been suggested that the *galería* system was invented by indigenous peoples in the Tehuacán region. One hypothesis is that as springs dried up on the valley floor, the inhabitants of the valley dug tunnels into the hillsides from the base of the spring in order to increase the flow of water. Once the technique was discovered to be successful, it was transferred to other areas (officials of the Tehuacán SRH office, in Cleek 1972:14). Although we have not seen wells of this type, descriptions of the springs in 1924 do declare that "the water of all springs flows from shafts . . . extended backwards [into

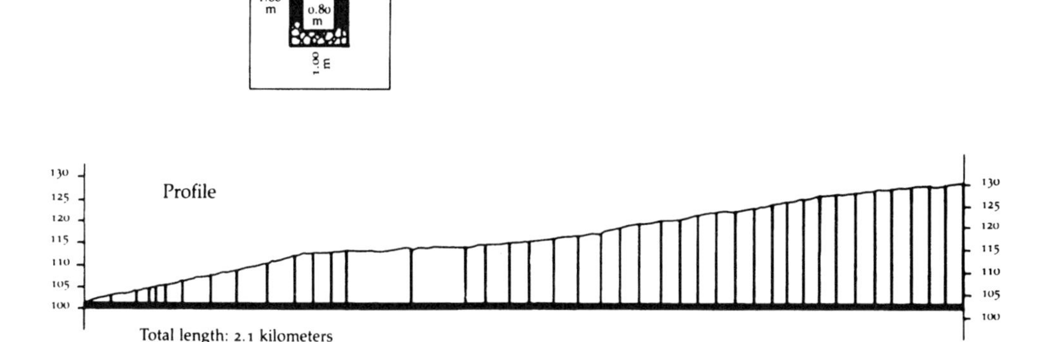

Figure 5. Galería Filtrante San Gabriel (after a figure supplied by Comisión del Papaloapan, 1977)

the] slopes" (Paredes Colín 1960:117, quoted in Cleek 1972:14).[9] Woodbury and Neely's work during the 1960s produced no evidence that the *galerías* were of pre-Hispanic origin, but they acknowledge that lack of time did not permit a more complete investigation (1972: 147–149). Alberto Rousel Castro (1942) and Jeannette E. Sherbondy (1982), however, have claimed that a *galería*-like technology originated independently in the New World along the coast of Peru and that the technology may have diffused north to Mexico.

Others have argued that the *galería* system was brought to Mexico by the Spanish, who learned the technology from the Moors during their occupation of Spain. In the Old World, the *galerías,* called *qanats,* have played an important role in the settlement and development of agriculture in arid regions. According to Henri Goblot (1979), the technology itself originated in Persia around 1500 B.C., as a system to drain mines, and later spread to Egypt, the Levant, and Arabia in Achaemenid times (550–331 B.C.). The Arabs carried the technology of *qanats* across North Africa into Cyprus and Spain; they are also found in Central Asia, Western China, and on a more limited scale in dry regions of Latin America (English 1968:170). The extension of the *qanat* system seems to have followed Islamic trade routes into China.

Despite the extensive use of irrigation in many regions of Latin America, *galería* systems are rare. Only in the Puebla-Tlaxcala region has the system been developed intensively by the local population (Seele 1969:3).[10] It is possible that the *galerías* were introduced to the valley by Spanish priests or by the owners of large haciendas or ranches who were attempting to develop independent sources of water. Laborers who were hired to do the construction, once having learned the techniques, could have moved on to form associations to generate capital or hire additional labor to develop their own *galerías*. The organizational structure of the *galerías* is similar to that of the barrio associations and the canal associations. We will discuss this in greater detail in Chapter 5.

Although several *galerías* trace their origins to before the beginning of the twentieth century, including Galería Sociedad Agrícola de Chilac, the number of *galerías* grew slowly during the 1920s. The period of greatest growth was between 1944 and 1969, when the number grew from 42 to 130. During the 1960s, more than 80 new *galerías* were constructed, despite the efforts of the SRH to control the construction in many areas in the valley. The *zona de veda* (restricted construction area) was instituted in 1946 because of a concern that new *galerías* would affect the flow of water in existing *galerías*. There was growing conflict between members of water associations. New associations built close to existing systems and there was strong pressure from current users to restrict the building of new *galerías*. The period of greatest construction corresponds to both increased agriculture in the valley and population growth. The rapid expansion of commercial agriculture during this period was definitely a major factor stimulating the growth of *galería* systems.

The other major source of water in the region is the private- and government-financed *pozos profundos* (deep wells) that are being constructed throughout the valley. These *pozos* are extremely expensive to build, requiring a large capital investment and highly trained personnel to do the drilling. By 1982, there were sixty-four pumps in the valley irrigating a total area of 1,600 hectares (SARH 1982:28).

Because of the total cost and the fact that payment cannot be spread out over time, only wealthy individuals can afford to drill without government aid. The wells are regarded by many campesinos as a dream for the future, but they are currently beyond their means. At the same time, many people fear that the wells will lower the water table and reduce the flow of water in the *galerías filtrantes*, a serious concern for those who own shares in them. The wells and pumps represent a new technology that is controlled by one sector of the population, a sector that is non-Indian and urban-based. The wells are viewed by speculators as an opportunity to make money by selling the water. We will return to this process later in the book.

All four forms of water acquisition are, in theory, under the control of the Secretaría de Agricultura y Recursos Hidráulicos (SARH). With the exception of a temporary ban during the 1960s on groundwater use, which was violated by powerful political figures, the government policy has been one of benign neglect.

The productive valley sustains a relatively high population density, 94.7 persons per square kilometer, and is currently experiencing significant population growth (between 1960 and 1980, the growth rate was 3.4 percent per year). Therefore, most of the people in the region are young, with 70 percent of the population below thirty years of age (SARH 1982:29). The major towns, communities with populations of over six thousand, have been the source of most of the population expansion. The *municipios* of the southern valley have grown rapidly in the last decade. Between 1960 and 1970, Ajalpan grew 38 percent, Altepexi 49 percent, Coxcatlan 23.6 percent, San José Miahuatlán 41.6 percent, and Zinacatepec 66 percent. Only San Gabriel Chilac did not grow significantly.

Table 2 shows the populations in four *municipios* from 1930 to 1980. Demographic expansion has been steady since 1930, but the rate of increase was greatest during the seventies; the trend is continuing in the 1980s.

The land area of the four *municipios* is shown in Table 3. In Altepexi the amount of land cultivated remained relatively stable up to 1970. During the seventies the *municipio* put an additional 950 hectares under the plow; the other three *municipios* underwent similar increases. It should be noted that the amount of land under cultivation in any one year varies as a function of how much water is available for irrigation. Chilac, for instance, had a water crisis in the late sixties as reflected in the reduction of cultivation area in 1970. As a general trend for the seventies, the increases in population, the construction of new irrigation systems, and the cultivation of more land are clearly evident with no signs of abating.

AGRICULTURE

Agriculture in the Tehuacán Valley is divided into two major types: mountain farming of lands above 1,800 meters in the oak and pine forests of the surrounding mountains, and valley farming, which includes the foothills and the valley floor.

The mountain farms are located on ground that may be sloped as much as forty-five degrees. The fields are plowed to follow the

Table 2. ***Population in Four* Municipios*, 1930–1980***

Population	*1930*	*1940 (% Change)*	*1950 (% Change)*	*1960 (% Change)*	*1970 (% Change)*	*1980 (% Change)*
Ajalpan	9,903	7,700 (−22)	10,018 (+30)	15,638 (+56)	21,568 (+38)	34,285 (+59)
Altepexi	2,534	2,771 (+9)	4,037 (+46)	4,467 (+11)	6,661 (+49)	12,521 (+88)
Chilac	6,659	6,868 (+3)	6,830 (−0.5)	7,255 (+6)	7,443 (+3)	15,330 (+106)
Zinacatepec	3,909	4,131 (+6)	4,545 (+10)	4,400 (−3)	7,310 (+66)	17,000 (+133)
Total	23,005	21,470 (−7)	25,430 (+18)	31,760 (+25)	42,982 (+35)	79,136 (+84)

Source: Secretaría de Programación y Presupuesto 1982*b*.

Table 3. ***Agricultural Land Use in Four* Municipios*, 1950–1980***

Municipio	*Total Area in Hectares*	*Potential Land for Use*	*(% of Total Area)*	*Area Cultivated in Hectares*			
				1950 (% Used)	*1960 (% Used)*	*1970 (% Used)*	*1980 (% Used)*
Ajalpan	38,400	9,676	(25%)	7,441 (77)	3,642 (38)	5,072 (52)	6,780 (70)
Altepexi	6,400	3,591	(56%)	1,529 (43)	1,908 (53)	1,601 (45)	2,554 (71)
Chilac	14,100	8,339	(59%)	1,232 (15)	2,287 (27)	618 (7)	4,160 (50)
Zinacatepec	8,600	7,075	(82%)	1,325 (19)	1,047 (15)	1,484 (21)	2,212 (31)
Total	67,500 ha.	28,681	(42%)	11,527 (40)	8,884 (31)	8,775 (31)	15,706 (55)

Sources: Comisión del Papaloapan 1983; Secretaría de Programación y Presupuesto 1982*b*.

Notes: Column 1 is the total number of hectares per *municipio*, both rural and urban. Column 2 is the total number of hectares of land with agricultural potential as classified by SARH.

contour of the slopes and are broken by earth dikes that rise ninety to one hundred centimeters above the furrow contours. The purpose of the dikes is to stop erosion and crop damage from water that may break through the furrows. The only source of water is rain, the quantity and regularity of which is greater than in the valley, especially in the Sierra Zongólica. Single crops of corn, beans, and squash are planted in April and harvested in late September or early October. The fields can be cultivated for only two to three years at a time, because the nutrients leach rapidly out of the soil on these steep slopes. The fields then lie fallow for up to five years before they are cultivated again.

Valley farming is more varied and intensive than mountain farming. A generally arid appearance and the predominance of cacti and shrubs give the impression that not much can grow in the valley. To the contrary, it is very intensively cultivated. Corn was cultivated early in the valley's prehistory and continues to be the basis of subsistence and capitalist agriculture. Today, the campesinos grow *elote,* or green corn, eaten as corn on the cob, as a principal cash crop. In order to harvest at the end of the dry season, when the prices in the urban markets are the highest, all of the green corn is grown on irrigated land.

Because of the mild climate, 2.8 crops of corn can be grown in one year. Some of the corn is allowed to mature and dry, the exact amount depending on the price that is being paid for green corn in Mexico City, Orizaba, Tepeaca, and Veracruz. The price of green corn fluctuates rapidly, and it was these rapid fluctuations that influenced people to install the first telephones in the valley, so that they could keep track of the prices for produce in the major urban markets. When the prices are low, campesinos will let their corn mature; when it is dry, they use it for tortillas and other corn-based foods. Along with beans, *ayocote* (a larger bean), and vegetables, corn is the basic subsistence crop. Pigs, chickens, goats, and occasional wild fruits make up the balance of the diet.

Corn is rotated with *jitomate,* tomato, and other vegetables at least once a year. Vegetables have recently become a major commercial crop in the valley. Garlic, the most important commercial crop in Chilac, is sold by vendors all over Mexico. Other crops are alfalfa and sorghum, which are used for animal or chicken feed, dry peppers, peas, and sugarcane. Saffron is also cultivated and exported to the United States.

Sugarcane grows mainly in the southern end of the valley, where there is less threat of frost and water is more plentiful. Sugar plantations and mills have played an important role in the history of the valley. Not only was sugar the major crop grown by the haciendas that once dominated the valley, but sugarcane cutting was and still is an important source of income for many families.

The soils in the Tehuacán Valley are either intrazonal, ranging from halomorphic alkaline and saline to calcimorphic rendzina soils, or zonal sierozem, characteristic of arid and semiarid regions. The parent material in the rocks that border the valley has been responsible for the large concentration of calcium and carbonate compounds in the soil. Alluvial soils have originated from water-transported deposits that were laid down during earlier geological stages (Geo-Re 1981; Stevens 1964:278; Thomas 1968:8).

The soil of the valley has remained fertile over centuries of irrigation through the constant addition of manure and of alluvia brought from riverbanks. Eventually these soils may become unproductive unless the excess calcium salts and carbonates are removed by the leaching action of surface water or by manual means.

Three types of land are cultivated in the valley: the limited amount of alluvial soil located in the canyons cut into the lower slopes of the Sierra Zongólica, along parts of the Río Zapotitlán, and in the area where the Río Salado joins the Río Zapotitlán; marginal land on the hillsides where dry farming techniques are used; and, by far the most extensive in terms of land use and output, the irrigated areas of the nearly level valley floor and the less steep hillsides. Alluvial canyon farming has been made possible by the deposit, over the centuries, of fertile soils from runoff and river drainage from the mountains. The moisture content in this soil is high during the entire rainy season and can support a single crop without irrigation. Fields are contour-plowed along the alluvial slopes for the construction of water-control devices in the same manner as on the mountain slopes. Corn and beans are the principal crops.

Dry farming is used over wide and scattered areas of the Tehuacán Valley. It is the most risky farming method, because the moisture content of the soil is negligible, and therefore crops need adequate amounts of precipitation. Dry farming is *limited* to the most marginal areas located on rocky and steep hillsides, which are unsuited to irrigation. Rocks are used to make low terraces to prevent the rapid runoff of rainwater and to retain the scarce soil moisture for as long as possible. There is also a system of contour plowing and the use of dikes for water control. The main crop is corn, because there is not enough moisture to support a simultaneous planting of beans. The growing season extends from April to September and the yields are extremely low. Lack of rain can often ruin an entire crop, and in a good year a yield of 300 kilos per hectare is considered average. An equivalent yield for irrigated corn is 1,600 to 2,000 kilos per hectare (weight yields are of dried grain that has been removed from the cob).

The third and most important type of agriculture is carried out by means of irrigation. The technology of how, and in what amount, irrigation water is obtained was described earlier. Here we shall discuss the cycles and techniques of irrigation agriculture, using data from the *municipio* of Altepexi. This agricultural cycle is typical of most of the *municipios* in the valley, although there are local variations in the timing of actual planting, the application of irrigation water, and other related tasks.

Double cropping is practiced by agriculturalists who have access to a year-round supply of water. Single cropping is practiced by those who have no irrigation water of their own and must purchase water from others, and by part-time farmers who engage in other activities, such as wage labor, basket making, or pottery. The first cycle begins in January and ends in late July or early August. The second cycle begins immediately after the first harvest and ends in January. There is some variation in planting and harvest times, and the January-August-January cycle is the most typical that we observed in 1972, 1973, and 1975. The most significant variation we observed was that some farmers planted at the end of February or the beginning of March. The monthly schedule for the distribution of irrigation water determines the exact timing of all other agricultural tasks, resulting in a thirty-day range of variation.

The first cycle begins with the initial deep plowing (*barbecho*) and subsequent soaking (*remojo*) of the soil. Soaking must be done with adequate amounts of water in order to reduce the salinity and alkalinity (pH) to a tolerable level. This first plowing is done using a steel plow, most frequently pulled by a pair of oxen. In some cases, a tractor is used. A wooden plow with a steel tip is also used occasionally, but the steel plow is considered best for turning the hard ground to a depth of twenty-five to thirty-five centimeters. The depth is important in order to achieve satisfactory soaking, leaching, and moisturizing during *remojo*. Six to eight days after the initial soaking, the fields are harrowed to break up large clumps, smooth out the surface, and evenly distribute the moist soil. The harrowing is generally done at a ninety-degree angle to the first plowing, because the application of water causes the surface to harden still more after baking in the sun for seven to eight days. Farmers say that it is best to plow before the ground is too hard. If the field is long and narrow, the second plowing follows the course of the first. A week later the field is plowed again (*surcado*) to make deep furrows for the planting that follows directly behind the furrow plow. The furrows are twenty to thirty centimeters deep and are placed about one meter apart. Planting (*sembrar*) is done by using a shovel to make a ten-centimeter hole in the furrow into which are tossed three to four corn kernels. If beans are planted simultaneously with corn, two beans are added to the same hole. These holes are spaced ninety to one hundred centimeters apart, along the furrow.

Chemical fertilizer, when used, is applied prior to seeding, and the third plowing will also serve to mix fertilizer with the soil. The use of chemical fertilizer is not very widespread, and many farmers rely exclusively on the abundant and much cheaper animal manure. The large commercial chicken industry near Tehuacán is a major source of natural fertilizer. Chicken droppings and cow manure are mixed with an equal part of soil and then spread along the furrows after the seeds have been planted. Some farmers apply the animal fertilizer before furrowing, especially when it contains large amounts of lumpy and coarse refuse.

After the planting, normally completed by the middle of February, the "first irrigation" (*primer riego*), which is actually the second application of water, takes place one month after the initial soaking mentioned above. Prior to this first irrigation, however, the irrigation infrastructures (i.e., all the physical structures such as canals and dikes, which facilitate and control the application of water) must be inspected and repaired. The borders around the field and the level of the field itself hold up well from season to season and may require only minor repairs, such as patching or raising, if the tops wear down from the effects of rain, wind, and human transport. Low spots in the field may need to be filled in, and leveling is often necessary to assure the nearly perfect level that is necessary to minimize water loss and maximize even infiltration into the soil. Some campesinos choose to construct dikes that subdivide the field into smaller areas of different levels. This is frequently necessary when the land is uneven, making it impractical to bring the entire area to the same level.

As noted before, the field is irrigated ten to fifteen days after planting, which means that the seedlings have already sprouted from the soil and the farmer can spot gaps in the furrows where no plants have appeared. High-quality seeds will leave very few gaps. Some farmers will quickly replant with new seeds before the water is applied. The water is set to arrive at an appointed time and is directed into the field by opening a small gravel dam that leads from the feeder canal to the field. The water flows along the side of the field and fills the furrows one by one to a depth of ten to fifteen centimeters, and the porous, deep-plowed, dry soil absorbs the water very quickly. Depending on the volume of water, the irrigation will last up to twelve hours. Irrigation systems that have flow rates of fifty to eighty liters per second take twelve hours to irrigate areas of one hectare, while a large, high flow-rate system such as La Purísima No. 3 can irrigate the same area in three hours. After the flow of water stops, it takes only four to five hours for all the water to soak into the ground. After this first irrigation there will be three or four more at intervals of thirty days until the corn is harvested.

A week to ten days after the first irrigation the corn plants grow rapidly. The plants grow about fifteen to twenty leaves in the first month. During this time, the demand for water and nutrients, especially nitrogen, is extremely high. By the end of the first month the corn plants are forty to fifty centimeters high and well above the furrows. By the time the second irrigation takes place, they will be at least seventy-five centimeters high.

The second and subsequent irrigations will wash away soil from around the base of the plants, which often collapse, especially during the frequent high winds. The bases of the plants must, therefore, be shored up (*segundas*). This can be done by a simple adaptation of the plow or it can be done manually using only a shovel. Most campesinos prefer the manual method, but frequently there is neither time nor labor enough. When using the plow, a small wooden board is placed on the side of the point to throw the soil up against the plant stalks. Two passes of the plow, in opposite directions, are made down each row, throwing the soil up against the plants. The net result is that the plants will now be hilled with furrows between the rows. A yoke of oxen can hill half a hectare in one day, whereas by hand it takes three to four times as long.

The toppling over of corn plants is a serious problem in hard desert soils. Corn root systems are shallow and provide little support. The warm weather in the Tehuacán Valley is, in part, responsible for the growth of very tall and thin corn stalks so that ears may begin to appear two meters above the ground. These plants are, therefore, top-heavy. The critical toppling period is during and immediately after the third irrigation, because the ears are well toward reaching their mature weight. If the stalk does not fall at this time, chances are that it will not fall at all.

Corn over ten weeks old has shed its pollen, and ear development is proceeding very rapidly. The kernels are swelling with the white sugary liquid that is characteristic of green corn. The *municipio* of San Sebastián Zinacatepec, south of Altepexi, specializes in producing green corn in growing cycles slightly under four months long, which enables the production of three crops a year. Zinacatepec has better soils and more available water, which allow for triple cropping. The large amounts of chemical and animal fertilizer that are added to the soil are largely responsible for the maintenance of soil fertility.

The second, third, fourth, and fifth irrigations take place 45, 75, 105, and 135 days, respectively, after planting, and harvesting occurs at 150 to 160 days. The fifth irrigation is considered optional by many farmers. Those who want to plant a second crop immediately after the harvest of the first claim that the last irrigation functions more to keep the soil moist than to provide moisture for the corn that is now drying on the stalks.

Weeds are removed sporadically during the growing cycle. They are not a major problem in terms of competition for soil nutrients and moisture, as is the case in the more humid and low-lying areas of the states of Veracruz and Oaxaca. Today more insecticides and fungicides are being used, particularly on *jitomate* crops. One reason for increased use has been the availability of credit with terms that require the use of certain pesticides, fertilizers, and other inputs.

The time between the shoring up of the plants, the irrigations, and weeding is used to repair and improve the irrigation infrastructure. Each irrigation and the effects of both rain and wind combine to erode the borders along the fields and the canals that conduct the water to the furrows. These constantly need to be repaired in order to avoid water loss during irrigation. The repair and improvement process is informative because it constantly tells the farmer about the slope of the land, the ability of built-up terraced areas to hold water in place without runoff, and the maintenance of proper contours for slightly inclined fields. This information is used to make changes for the second crop.

About three weeks before harvest, some farmers double over the corn stalks so that the corn cobs hang waist high upside down. The rationale for this practice is that corn should not be harvested until it is dry, and the ears on doubled-over stalks dry much faster. The speed of drying, however, is a matter of controlled moisture loss from the kernels, which, if excessive, results in wrinkled and shrunken kernels. The upside-down position of the ear wraps the husk around the cob and seals the bottom, thus reducing excessive moisture loss and preventing rainwater from entering. When it is felt that the yield will be low and the corn will be of inferior quality, some farmers let the ears stand and dry in an upright position.

After the harvest and transport, farmers spread the corn on the floor of their patios, living areas, or any available flat surface. With the help of the entire family, using long wooden sticks, the kernels are clubbed from the cob. The kernels are then swept up and put into large sacks. The cobs are stored to be used for fuel in the kitchen fire. After the shelling, the kernels contain dirt and impurities that are sifted out using a metal screen attached to a wooden frame. The corn is now ready to be weighed for sale. Enough will also be put aside to meet family needs until the next harvest.

The yield of corn from one hectare is highly variable, depending on the condition of the soil, the effectiveness of irrigation, the amount of residual salt, the depth of the original plowing, and the amount and type of fertilizer used. The average yields in the *municipio* of Altepexi in 1973 were 1,700 kilograms of dry shelled corn from one hectare of irrigated land—the lowest yield was 950 kilograms and the highest was almost 2,000 kilograms. A family of six consumes approximately 1,200 kilograms per year, which means that those who cultivate more than one hectare can harvest crops that are of considerable commercial value.

Between 1973 and 1981, many of the agriculturalists in the region changed from predominantly subsistence to almost completely cash-crop cultivation. Of those who changed, most are now cultivating green corn. In previous years, agriculturalists from San Sebastián Zinacatepec were the only green corn cultivators in the valley and were reputed to have a monopoly on the marketing in Mexico City. Now, however, green corn is cultivated in Ajalpan, Altepexi, Chilac, Tepetzingo, and San Diego Chalma, as well as in Zinacatepec.

Another commercial crop that has increased in the region is *jitomate* (known as kitchen tomatoes in the United States), but it is not as widespread as green corn because of higher water requirements and the need for frequent pesticide applications. *Jitomate* cultivation is also more labor-intensive. Although the amount of labor and other inputs is higher, the return on *jitomates* is over double that of green corn. The wealthy campesinos switched to both green corn and *jitomate* cultivation during the 1960s, whereas others cultivated only green corn.

The reason for switching to green corn and *jitomate* becomes more apparent with a comparison of cost per hectare of cultivation, yields, and the wholesale market prices. For traditional maize cultivation using fertilizer, adequate water, and an "improved" variety of seed, the yield per hectare was reported to have been as high as 3,000 kilos in the late 1970s. If these are accurate estimates, the yields have doubled from the early 1970s. In 1978 the cost of production was 8,600 pesos per hectare, and the market value per kilo was 3 pesos. A yield of 3,000 kilos would therefore generate a profit of 400 pesos per hectare and a yield below 2,860 kilos would be a loss. Since the maize was used primarily for home consumption, the actual profit or loss is meaningless, because the labor was usually contributed by nonpaid family members. These cost calculations use the going rate for peon labor and other inputs.

Green corn, on the other hand, presents a different economic picture. The total cost per hectare for all inputs was 7,100 pesos in 1978 (the lower cost reflecting a shorter growing season requiring less water). The yield was fifty 100-kilogram sacks (*bultos*) of green corn, worth, on the average, 140 pesos per bulto, and 1,700 kilograms of shelled corn worth 3 pesos per kilo, which gave a gross income of 12,100 pesos per hectare. The profit was nearly 5,000 pesos, or about ten times more than maize. It should be noted that each agriculturalist left a small portion of his *milpa*, or cornfield, (approximately 5 percent) for the full six months in order to produce seeds for the next planting. Some agriculturalists in Altepexi said they were able to grow three crops a year, but the average in the region was five crops of green corn every two years.

Jitomate was even more profitable, but the inputs were extremely high. The total cost of inputs per hectare in 1978 was 19,000 pesos, and the yield was, on the average, 300 boxes, weighing 30 kilos each and having a market value of 100 pesos per box. The profit per hectare was thus 11,000 pesos. The high production cost of *jitomate* was due to increased water requirements, additional furrowing, and the high cost of pesticides. Consequently, only the wealthy campesinos had the financial resources and water to grow large quantities. Those who did were able to accumulate large profits, which were confirmed by new trucks and other expensive consumer items.

By the summer of 1981, nearly all the agriculturalists in the valley were cultivating green corn. The market price in Mexico City had gone as high as 500 pesos per sack while inputs had only increased slightly from the 1978 figures listed above. The price of *jitomate* had also increased to over 300 pesos per box, and many were trying to cultivate as many hectares as available water would permit.

It is very difficult to determine the relationships between the quantity of water that is applied to a field, the amount needed by the plant for maximum crop yield, the frequency at which water must be applied to maintain proper soil moisture, and the particular environmental circumstances of an area that affect irrigation. Furthermore, the actual amount needed by the plant for metabolism and transpiration and the amount needed for a maximum crop yield may be different. The reason for this apparent discrepancy is that in areas with high mineral and salt concentrations in both the soil and water, more water must be applied to the field than is necessary to support optimal plant growth under normal circumstances. The extra water is necessary to maintain soil fertility, because inadequate amounts of irrigation water will evaporate very quickly before the water can soak in and will leave visible deposits of salt and carbonate crystals. The next irrigation, if also inadequate, will add to the salt on the soil. As the alkalinity increases, the ion exchange capacity of the soil decreases, thus retarding the absorption of nutrients by the root systems. The high salt, mineral, and carbonate concentrations in the Tehuacán Valley are the most important limiting factors influencing the maintenance of soil fertility and continuous cropping over the years.

The problem with discussing adequate versus inadequate amounts of irrigation is to define exactly how much water is needed to irrigate a particular parcel of land. There are numerous variables, such as crop type, soil conditions, climate, and altitude, that further complicate the problem. Many agronomists, farmers, and other knowledgeable persons state that enough irrigation water has been applied if the soil remains moist during the entire growing cycle. We know of no studies that have quantified the amount of water needed for maize cultivation in the Tehuacán Valley or in any area that is environmentally similar.

Woodbury and Neely (1972:99) cite some figures for the water requirements for sugarcane in the Río Salado subarea of the Papaloapan Basin. It is generally accepted that cane needs more water than corn, but it may be useful to compare these quantities with the amounts of water normally used to irrigate a field of corn in the *municipio* of Altepexi. The estimate for sugarcane is ten thousand cubic meters of water needed to irrigate one hectare. This figure is the total amount needed to produce the crop, but there is no breakdown as to how much is applied at what intervals. It must be assumed that all of this water is not applied at the same time.

The *galería* system Guadalupana, in Altepexi, produces a flow of approximately 75 liters per second, or 0.075 cubic meters per second. A farmer irrigating approximately one hectare applies the water for twelve hours, totaling 3,240 cubic meters per hectare. This amounts to 16,200 cubic meters for the growing season. The conclusion is that the corn farmer is applying 62 percent more water than is needed for an equivalent crop of sugarcane, which supports the argument that more water is placed on the corn fields than is necessary.

Both farmers and agronomists in the area agree that the amount of extra water is needed to reduce the buildup of excessive salt and carbonate concentrations. The need for "overwatering" prevents agriculturalists from placing more land under cultivation, but the fertility of the land already under cultivation can be maintained for longer periods. However, no matter how much water is applied, the fields eventually become impregnated with excessive amounts of salt and have to be abandoned. Woodbury and Neely (1972:135) state that, since calcium carbonate dissolves with relative ease, if fields are allowed to lie fallow for several years, the light rains in the valley might slowly flush away the deposited *tepetate* and cleanse the fields. If this is true, the carbonate-saturated soils are renewable. In the *municipios* of Chilac and Altepexi, however, there are large expanses of saline land where virtually nothing grows. Local farmers say that these lands have been this way for "many years" and will probably stay that way indefinitely. C. Earle Smith (1965*a*:22) claims that a certain amount of deposited minerals and carbonates may be leached from the fields, but generally the process is irreversible. Once many salts have precipitated, particularly calcium salts, they are extremely difficult to redissolve. The resource base for agriculture is becoming degraded and depleted as the result of farming practices currently in use.

The number of *galería* systems currently in operation most certainly withdraws water faster than it is replaced by infiltration, and many systems are close enough together to affect each other's flow rate. Farming in the mountains has removed most of the forest cover with the inevitable results of rapid runoff, decreased infiltration for aquifer recharge, and flash floods that can damage fields at lower altitudes. As confirmed by the need to make *galería* tunnels deeper, the water level in the first stratum is decreasing. Even though the increased *galería* construction of the last twenty years, deforestation, and extensive droughts have all contributed to the lower water level, the situation is potentially more complex.

The extensive proliferation of *pozos profundos* raises a number of important questions. How large are the deep-water reserves? Is there a possible connection between the two water-bearing strata? If there is, the upper stratum could drain into the lower as it becomes depleted and put many *galería* systems out of operation. The government and private enterprises have ambitious plans for sinking a large number of new wells all over the southeastern valley to increase irrigated areas. There are also plans to establish new industries for fertilizer production and canneries for agricultural produce, such as green corn and *jitomate*; all require large amounts of water. An existing cooking oil factory in San Pablo Tepetzingo competes with agriculture for water and is a menacing source of environmental pollution.

Another series of studies completed in 1983 by the Papaloapan Commission on the changes in flow rates of *manantiales, galerías,* and *pozos profundos* show some disturbing trends. The flow rates of nine

Table 4. *Changes in Flow Rates of Natural Springs*

Spring	Municipio	*1976 Lps*	*1983 Lps*	*Change Lps*	*% Change*
North of Tehuacán					
Sin Nombre	Tehuacán	77.0	25	−52.0	−67.5
San José	Tehuacán	100.0	90	−10.0	−10.0
San Jorge	Tehuacán	99.6	90	−9.6	−9.6
San Miguel	Tehuacán	84.4	76	−8.4	−9.9
Santa Ana	Santiago Miahuatlán	275.6	248	−27.6	−10.0
South of Tehuacán					
La Taza	Chilac	591.6	450	−141.6	−23.9
La Ciénaga	San José Miahuatlán	616.0	556	−60.0	−9.7
El Coyoatl	San José Miahuatlán	43.6	39	−4.6	−10.6
El Tochatl	San Sebastián Zinacatepec	121.5	109	−12.5	−10.3

Source: Comisión del Papaloapan 1983.
Lps: liters per second.

natural springs were monitored over seven years, 1976–1983; the results are shown in Table 4. An 11.8 percent change was observed as an average from all nine *manantiales,* and when subdivded into north and south of the city of Tehuacán, the decreases were 9.9 and 13.6 percent, respectively.

Twenty-one *galería* systems were also monitored over the same approximate time period; the average decrease, as shown in Table 5, was 38.2 percent. As with the *manantiales,* there was a greater decrease to the south of Tehuacán: 24.8 versus 46.1 percent.

The measurements of the new *pozos profundos* (Table 6) also show a downward trend but with the largest decrease to the north of Tehuacán, confirming the earlier geological findings locating deep aquifers in the southern valley (Comisión del Papaloapan 1983). Anecdotal accounts by farmers about decreases in water flow are numerous. These can often be explained by poor maintenance or the idiosyncrasies of a particular aquifer, but this is the first time that flow rates have been accurately measured over long time periods. Popular accounts have been confirmed by "scientific" findings.

The groundwater resource assumed by most local residents to be limitless has, in fact, shown itself to be finite. Up until the release of these findings, there was no hard evidence showing general and consistent flow-rate trends in multiple systems from three distinct

Table 5. ***Changes in Flow Rates of* Galerías Filtrantes**

Galería	Municipio	*Lps*	*Year*	*1983 Lps*	*Change Lps*	*% Change*
North of Tehuacán						
Cipiapa	Tepanco	36.4	1976	26	−10.4	−28.6
Hidalgo	Santiago Miahuatlán	229.8	1976	217	−12.8	−5.6
Benito Juárez	Santiago Miahuatlán	81.5	1976	53	−28.5	−35.0
Antonio Pailles	Chapulco	75.0	1976	69	−6.0	−8.0
La Providencia	Chapulco	18.3	1976	10	−8.3	−45.4
La Colectiva	Chapulco	38.4	1976	26	−12.4	−32.3
El Mano de la Providencia	Tehuacán	93.9	1976	72	−21.9	−23.3
Calavera & Cozahuico	Tehuacán	143.7	1968	56	−87.7	−61.0
El Huizache	Tehuacán	178.6	1976	143	−35.6	−19.9
South of Tehuacán						
San Sebastián de los Arcos	Ajalpan	27.2	1976	8	−19.2	−70.6
San Angelo Pantzingo	Ajalpan	46.5	1976	44	−2.5	−5.4
El Carmen	Altepexi	106.0	1976	50	−56.0	−52.8
San Pedro	Altepexi	13.7	1976	12	−1.7	−12.4
San Rafael	Altepexi	62.2	1976	50	−12.2	−19.6
Corazón de Jesús	Zinacatepec	17.0	1968	12	−5.0	−29.4
El Nilo	Zinacatepec	98.7	1976	6	−92.7	−93.9
Hidalgo	Zinacatepec	41.6	1976	12	−29.6	−71.2
San Juan	Zinacatepec	127.0	1968	59	−68.0	−53.5
Mercedes de María	Chilac	52.6	1976	19	−33.6	−63.9
Agrícola de Chilac	Chilac	11.4	1976	9	−2.4	−21.1
San Gabriel	Chilac	65.0	1976	32	−33.0	−50.2

Source: Comisión del Papaloapan 1983.
Lps: liters per second.

Table 6. ***Changes in Flow Rates of* Pozos Profundos**

Pozo	Municipio	*Lps*	*Year*	*1983 Lps*	*Change/ Year*	*% Change/ Year*	*Total Change*	*% Total Change*
North of Tehuacán								
San Diego Chalma	Tehuacán	27.5	1976	22	−0.71	−2.6	−5.0	−18.2
San Vicente	Tehuacán	50.0	1976	33	−2.4	−4.9	−17.0	−34.0
Guadalupana	Chapulco	60.0	1970	52	−0.61	−1.02	−8.0	−13.3
Providencia	Chapulco	25.5	1976	0	−3.6	−14.3	−25.5	−100.0
Emiliano Zapata	Miahuatlán	30.0	1981	0	0.0	0.0	−30.0	0.0
South of Tehuacán								
Col. Vista Hermosa	Chilac	9.0	1982	5	−4.0	−44.4	−4.0	−44.4
Gómez Román	Chilac	53.0	1976	75	+3.1	+6.0	+22.0	+42.0
San Rafael 1	Altepexi	82.0	1981	82	0.0	0.0	0	0.0
San Rafael 2	Altepexi	82.0	1981	82	0.0	0.0	0	0.0
Pantzingo	Ajalpan	45.0	1981	37	−4.0	−8.9	−8.0	−17.8
Venta Negra	Ajalpan	50.0	1981	41	−4.5	−9.0	−9.0	−18.0
San Martín de P.	Zinacatepec	80.0	1981	76	−2.0	−2.5	−4.0	−5.0
San Sebastián	Zinacatepec	90.0	1981	86	−2.0	−2.2	−4.0	−4.4
San Isidro	Zinacatepec	35.0	1981	30	−2.5	−7.2	−5.0	−14.3

Source: Comisión del Papaloapan 1983.
Lps: liters per second.

sources: *manantiales, galerías,* and *pozos profundos.* The parallel decreases in both *manantial* and *galería* flows to the south of Tehuacán are convincing evidence that current use greatly exceeds recharge rates. The measurements obviously are not the result of low precipitation in any one year or agricultural season. From an ecological point of view, continued water use at present levels will eventually force a reduction in the number of irrigated hectares and result in drastic decreases in agricultural output. When and how quickly this will happen is open to debate; an educated guess would be by the beginning of the 1990s.

It is in this social and physical environment that traditional agriculture continues to function. Local indigenous irrigators continue to have control over *galería* water, but competition for water promises to increase as outside capital and technology drive wells deeper into the valley floor. Current conflicts between individuals, irrigation associations, private interests, and the government are numerous, and the prospect for more conflict and violence is even greater.

In San José Miahuatlán conflicts between an association of irrigators and the *municipio* government erupted into violence in early 1982. Many of the *municipio* officers were killed because they were trying to enforce a government decree allocating a portion of *galería* flow to domestic use in the town. Government personnel withdrew, and plans have been made to meet domestic needs with water from new *pozos profundos.* Until that time, there will be a continuing and acute water shortage in the *municipio.*

It is within this framework that we will examine and analyze individuals and groups as they compete for control over means of production, manage resources, and resolve conflicts. In order to have a better perspective on the current situation, in the next chapter we return to the prehistoric past and look at the use of the same resources fought over today. We then trace the development of technological, economic, political, and social organization through the Conquest, colonial period, independence, and revolution up to the multiple sociopolitical institutions that shape current complexities in a fragile environment.

3. THE PRE-CONQUEST DEVELOPMENT OF AGRICULTURE AND IRRIGATION

THE RELATIONSHIPS between environment, resources, technology, and sociopolitical organization are complex. In modern society it is nearly impossible to sort out the links, covariations, cause-and-effect relations, and neutral factors. Anthropologists have traditionally examined less-developed societies before tackling current complexities. Data from the past and present in the Tehuacán Valley provide vital links between the two.

Historical continuity not only serves to reconstruct past events, but also provides a basis for interpreting the present. The current complexities of the social organization in the Tehuacán Valley have roots that go deeply into the historical and prehistorical past. We are particularly fortunate in having a well-documented record going back to at least 10,000 B.C. The record consists of archaeological sources and descriptions of the conditions just preceding the Spanish Conquest. The purpose of this chapter is to outline the principal events and changing conditions throughout this long sequence to better understand the current relationships between individuals, groups, institutions, and natural resources.

The chapter examines the evolution of increasingly complex food production systems, ranging from hunting and gathering to fully developed pre-Hispanic irrigation agriculture. Using published archaeological and ethnohistorical data and analysis, we discuss the relationships between levels of complexity in social organization, technology, and irrigation. It appears that every society with its unique ecosystems and adaptation must be carefully studied; broad generalizations often gloss over important details in local variations that can account for the emergence of specific structures and organizational variants.

The principal reason for including an archaeological and ethnohistorical chapter in what is essentially an ethnography of twentieth-century Tehuacán is to argue that important variables in the shaping of the earliest of organizations and politics continue to affect modern responses to the use of technology and resources. The present is but one section of a continuum of behaviors to cope with changing physical and social environments.

ARCHAEOLOGICAL CHRONOLOGY

The prehistory from human occupation to initial settlement to the Spanish Conquest has been reconstructed and interpreted by a large interdisciplinary team forming the Tehuacán Archeological-Botanical Project, sponsored by the R. S. Peabody Foundation for Archeology. The project gathered field data during the early 1960s and has since published five volumes, with a sixth currently in preparation. The theoretical and methodological innovations used by this team have produced a wealth of documentation and careful interpretations, using extensive comparison between past and present.

The object of the Tehuacán Project was to investigate the processes of long-term culture change, focusing on the origins of food production and the emergence of civilization. Artifacts and ecofacts were viewed as products of a cultural system interrelated with a wider ecosystem. The cultural system has been divided into the following three subsystems:

1. Adaptation to the natural environment—subsistence, technology, settlement patterns, and exchange.
2. Interrelations and adaptation of individuals—social system, kinship, economic relations, politics, rank, and class.
3. Ethos system—system of values, humankind's relations to the cosmos.

A cultural system is considered to be open, and significant changes in the subsystems have resulted in the formation of new relationships. Two basic conditions may cause changes: those that trigger or stimulate an event (sufficient conditions or sufficient causes); and a set of conditions allowing triggering, which by themselves would not cause change from one cultural system to another (necessary conditions or necessary causes). If one has a documented sequence of cultural systems and related subsystems in the Tehuacán Valley with plausible relationships to ecosystems and other culture sequences in Mesoamerica, one can make hypotheses about changes in subsystems that caused one phase to evolve into another (MacNeish and Nelken-Terner 1972:3–8).

The Tehuacán Project was truly interdisciplinary and included geological, hydrological, botanical, zoological, and traditional archaeological analyses. Based on these data, Richard S. MacNeish isolated nine cultural phases, spanning the time from first human occupation of the valley (ca. 10,000 B.C.) until the Spanish Conquest (A.D. 1521). The brief summaries are drawn from Volume 5 of *The Prehistory of the Tehuacan Valley* (MacNeish 1972:496–504).

Ajuerado Phase (10,000–7000 B.C.)

Based on a small number of randomly distributed sites, MacNeish suggested that the earliest occupation of the valley was located in microband camps consisting of two to three nuclear families. They were highly nomadic and occupation was for a single season or less. Hunting and trapping were the prime subsistence activities during all seasons, but some leaf and seed collecting and fruit and pod picking were also done during certain seasons. There is some evidence for scheduling, which implies a sequence of campsite selection based on seasonal variation in the availability of food. During the driest seasons the camps were closer to water. The population in the region during this three-thousand-year phase was on the order of one to five families divided into groups and has been classified as a nomadic microband community pattern. There are no contemporary analogies (MacNeish 1972:497).

El Riego Phase (7000–5000 B.C.)

By this time the settlement pattern shows a well-scheduled system of seasonality in which dry-season microbands fused to form macrobands during lusher seasons. The dry season was for hunting and trapping, and spring and fall were the time for seed collection and fruit picking. These seasonal food strategies were "satisfying" in that they consisted of a safe and reliable way to collect food. The population size was about 100 to 150. The changes in subsistence accompanied a more advanced technology, and the fission/fusion of the bands meant a more elaborate social organization as well as new ceremonial manifestations. This phase has been classified as having a pattern of seasonal micro-macro bands (MacNeish 1972:497).

An important question was to determine whether population pressure and subsequent settlement patterns caused the changes. Population pressure was probably not the major factor, because the increase was in the late El Riego Phase, after the major changes had occurred. It is more likely that population increases and settlement pattern changes were the result of environmental changes, a general desiccation of the valley resulting in less game. Increased scheduling was therefore a form of negative feedback with changing use of subsistence options such as an increase by seasonal reliance on vegetable food and less on hunting and trapping (MacNeish 1972:498).

Coxcatlan Phase (5000–3400 B.C.)

Here we see more macro- than microband encampments, with a slight population increase, but the settlement pattern is similar to the previous phase. The subsistence activities included some incipient agriculture, and these Tehuacán Valley sites are among the earliest examples of agriculture in the New World. The increase in food that resulted from domestication allowed for longer sedentary occupations in both small and larger camps. The microbands fused during the spring and summer months to collect plants. The domesticates planted in early spring were ready for harvest when seasonal wild plants were gone, allowing additional sedentary time, but there was not enough surplus for year-round residence. During the dry winter months the bands fissioned or the whole group moved to spots along the waterways on travertine and alluvial slopes. The camps were larger and represented a total population of one hundred to four hundred people. Population pressure may well have been responsible for the changes from the El Riego Phase (MacNeish 1972:498).

Seasonal scheduling during El Riego times apparently led to the development of the Coxcatlan subsistence patterns. The inhabitants of the late El Riego Phase must have returned to certain seed, fruit, or food collection areas on an annual basis. The seasonal clearing, weeding, and enrichment of the soil with human and animal refuse probably resulted in an unconscious selective pressure on the vegetation. The pressure could have resulted in a new and artificial environment that promoted or allowed for genetic changes in particular food species. This was horticulture: individual domesticates in limited areas or gardens. The additional food and longer occupation periods of particular sites led to larger populations and changes in settlement patterns toward macrobands. Larger and denser populations, in turn, led to increased dependence on horticulture, resulting in continued and spiraling changes during the Coxcatlan Phase with its principal pattern of macro-micro bands. New domesticates were introduced, and we see the first use of early domestic varieties of corn (MacNeish 1972:498–499).

Abejas Phase (3400–2300 B.C.)

Settlements along river terraces emerged as a new community pattern coexisting with many multiseason macroband camps and associated single-season microband camps. The emerging community pattern has been classified as central-based bands, consisting of people representing the full biological population from more than one home base who carried out a range of sociopolitical activities. The hamlets evolved because of increases in storable horticultural products and seed surpluses. Individuals continued to collect wild vegetation and hunt animals while planting as a supplement. During the winter dry season, the summer cultigen surpluses permitted the provisioning of sedentary hamlets, whereas those without adequate food were forced to fission into microbands. The Abejas Phase also marked a move from the previous "satisfying" strategy to a more optimizing strategy of striving to obtain the largest possible amount of food without considering potential dangers caused by unanticipated human or natural disturbances. In other words, more food could be produced but at a greater potential risk (MacNeish 1972:499).

Purrón Phase (2300–1500 B.C.)

Data are lacking from this phase to make inferences or speculations about sociopolitical organization (MacNeish 1972:499–500). However, settlements were semipermanent waterway communities.

Ajalpan Phase (1500–850 B.C.)

Not many sites have been found from this phase, and limited excavations provide for few generalizations. The settlement pattern consisted of semipermanent river terraces and associated macroband camps. These were more permanent and had a larger population than in earlier phases. Wet season *barranca* agriculture was a new and dominant food cultivation technique leading to an increased food surplus. This new technology is credited with being the key factor for the shift from the Abejas to Purrón to Ajalpan phases. Agriculture has been defined as the planting of cultigens in particular fields requiring preparation, maintenance, and very definite harvesting procedures.

More productive domestic corn hybrids were used during the Ajalpan Phase. There is also some evidence of cultural interaction with other regions, which may have caused changes in settlement patterns, population size, and density, rather than the reverse (MacNeish 1972:500).

Santa María Phase (850–150 B.C.)

The relationships between food production and population patterns were changing between 1200 and 850 B.C. Populations may have tripled during the early Santa María Phase, resulting in a nuclear village community pattern. Irrigation agriculture and orchard culture appear as new cultivation technologies, most probably because wet season *barranca* agriculture did not yield enough.[1] In comparison to seasonal agriculture, irrigation agriculture is a more efficient strategy that attempts to reach the highest possible level of food cultivation that can be maintained over a long period of time; however, the risks are higher, because of the constant threat of droughts and the destruction of irrigation infrastructures. The development of irrigation may have been the result of the exchange of ideas, produce, and artifacts with other Mesoamerican areas. In view of the available data, it is difficult to more exactly interpret the importance of outside influences. During this phase, urban civilizations were emerging in the neighboring Valley of Oaxaca and Basin of Mexico. Clearly, more complex socio-religious cultural systems were evolving (MacNeish 1972:500).

Palo Blanco Phase (150 B.C.–A.D. 700)

The data for this phase, primarily from site surveys and less from excavations, show greater cultural complexity and new types of settlement patterns. There are seven to eight settlement clusters, each cluster revolving around a single nucleated town (usually on a hilltop), and three are associated with secondary centers. Some may have salt evaporators and encampments. These towns probably functioned as administrative, political, and ceremonial centers, each town and village with its own internal hierarchical structure. The town centers and largest mounds contained the most elaborate plazas, and were surrounded by poorer architecture and smaller and smaller mounds. These were classified as advanced nucleated centers, and because of the ceremonial functions they are also called dioceses. A huge population increase from Santa María times, due to increased food supplies from larger-scale irrigation systems, has been interpreted as negative feedback between subsistence subsystem and population. Increasing nucleation and organization is seen as a response to administering the growing number and scale of water control systems. This is not to say that irrigation and population are solely responsible for the complex shift in social structure from Santa María to Palo Blanco, but they must have been significant factors (MacNeish 1972:500–501).

Venta Salada (A.D. 700–1521)

The transition to a new community pattern consisting of small city-states is not clearly understood. There are some ethnohistorical data for the late Venta Salada showing the emergence of a local elite that ruled these city-states, called *cacicazgos.* The cities of Teotitlán del Camino, Coxcatlán, Zapotitlán, and Mezontla are classified as capitals in the early Spanish documents. Some fortifications and other evidence for defense are widespread. Surface collections and preliminary excavations show, or strongly suggest, barrios of full-time specialists and two classes of structures—for nobles and for commoners. Within the surrounding state territories, there were clustered village subcommunities and hamlets. Many show specialized economic activities, producing, among other things, salt, stone tools, and mold-made pottery. Some village sites show evidence of specialized military activities, and others have shrines indicating specialized religious activities. There were further population increases, but there is no clear evidence from the Tehuacán Valley itself for the causes of these major sociopolitical changes. Important possibilities include trade relations, economic factors, militarism, urbanism, and other social factors (MacNeish 1972:501–504).

PREHISTORIC IRRIGATION SYSTEMS

In view of the arid climate, water must have been a constant concern since the beginning of agriculture in the valley, about 4000 B.C. The earliest cultivation efforts were timed to coincide with the rainy seasons and used the humid soil of deeply cut alluvial river valleys. Because of the many natural springs on the valley floor and the periodic flows of rivers, it is not surprising to find evidence of irrigation practices as early as 700 B.C. with the identification of the Purrón Dam complex. This was a large structure and could not have represented the first effort at irrigation. More modest beginnings could have been made much earlier during the three thousand years between the first cultivation and the first identifiable irrigation. A dam structure such as the Purrón could survive the millennia, whereas small diversion canals most probably could not.

In this section we describe these early irrigation systems based on the work of Woodbury and Neely (1972:81–153). Four types of irrigation are examined: the Purrón Dam complex, the Xiquila Aqueduct, the agricultural terraces and borders, and the large-scale canal networks associated with the principal natural springs. A fifth type, the contemporary chain well systems, or *galerías filtrantes,* will be the subject of detailed treatment in later chapters.

The Purrón Dam Complex

The Purrón Dam, the largest prehistoric water control structure encountered, is located in the southeastern part of the valley in the canyon of Arroyo Lencho Diego, 2.5 kilometers northwest of San José Tilapa. The canyon contains two dam structures—a small upstream structure and a larger dam that went through five stages of construction of increasing magnitude. The dams provided irrigation for alluvial fields to the southwest occupying an area of about 6.5 square kilometers. The main dam is located in a canyon constriction where the width is 400 meters and the cliffs on the sides are 40 to 50 meters high. Beyond this point the canyon widens and opens to the alluvial fields about 1 kilometer to the southwest. The present-day stream has cut through the dam and, together with erosion in the canyon, has provided data for the interpretation of five stages or levels of construction (Woodbury and Neely 1972:82–83).

The first level, constructed about 750–600 B.C., extended across only part of the 400-meter canyon, leaving a 175-meter gap. It was 6 meters wide, 2.8 meters high, and was composed of earth and small stones over larger rocks, faced with small to medium-sized stones. This catch basin covered an area 140 by 170 meters. There is evidence of silting behind the dam, demonstrating fifteen to twenty horizontal layers of alluvium, which represent a corresponding number of years before the silt made the dam useless (Woodbury and Neely 1972:83–86).

The second level, constructed about 600 B.C., was built over the first and measured 100 meters long by 3 meters high. The fill was divided into compartments separated by crude stone reinforcing walls of cobbles, and the exterior was faced with rectangular slabs. During this time, another dam was built a few hundred meters upstream from the main structure. It was a flat-topped structure measuring 550 meters long, 30 meters wide, and 3 to 5 meters high, and built like the second level of the main dam. On the flat top were three mounds that may have been the foundations for small rectangular houses. The upstream dam probably served as a coffer dam during the construction of the main dam. It may also have provided irrigation water for cultivation in the dry main dam reservoir during the winter months (Woodbury and Neely 1972:86–90).

The third level of construction, about 150 B.C. to A.D. 150, may just have been a repair of the top and western face of the main dam. There is no evidence of changes in the location of the spillways. The maximum area of the reservoir was about 400 by 600 meters. By the end of its use, a 3-meter alluvial buildup had occurred behind the dam (Woodbury and Neely 1972:90–91).

The fourth rebuilding, about A.D. 300, expanded the reservoir to a depth of 8 meters and an area of 400 by 700 meters. At this point the smaller dam upstream must have been covered, or nearly covered, with water. At the end of this period—the late Palo Blanco subphase—the dam and most of the sites in the canyon were abandoned (Woodbury and Neely 1972:91).

The fifth and final layer of construction may have consisted of structures that had nothing to do with the irrigation capacities of the dam. It was most probably done by people who were reinhabiting the canyon shortly after A.D. 1100 and who were using the dam as a building platform and not for irrigation (Woodbury and Neely 1972:91–92).

The first dam in the early Santa María Phase probably contained sufficient water for one corn crop per year, possibly two. The estimated volume for level two and subsequent levels is over eight times more and should have produced two, if not three, corn crops per year. During this time irrigated agriculture produced, in addition to corn, avocados, *zapotes*, and cotton. It is quite possible that the dam and increases in agricultural productivity are related to the larger number of occupied sites and the surge in population. But it is impossible to say whether the dam was responsible for the changes or whether the increased settlements had provided the organization, technology, and labor to construct the dam (Woodbury and Neely 1972:95).

Woodbury and Neely conclude that the use of irrigation agriculture most probably permitted population increases and the development of specialization (1972:95). It is difficult to determine why the dam was abandoned, but alluvial silt deposits in the reservoir were certainly a factor. Possibly there was a shift to other areas for farming, such as the nearby fields supplied by the Río Tilapa just to the southeast. Furthermore, the move may have been necessitated or even forced by the buildup of surface salts and hardpans on irrigated fields, thus drastically reducing yields; or other sociopolitical factors may have been the reason. Smaller dams have been found in river canyons, but none have approached the scale of the Purrón Dam complex (ibid.: 102).

The Xiquila Aqueduct

The Río Xiquila, draining part of the Mixteca Alta, runs into the Río Salado from the west at a point about 50 kilometers south of Tehuacán. Fifteen kilometers upstream from the Río Salado junction is an area called Las Huertas, a 2-kilometer stretch of fertile bottom land about 300 meters wide along the river. Black *zapote*, orange, lime, and banana orchard crops have been traditional here. Three *manantiales* bordering the river add 1,000 liters per second (lps) to the 100 lps river flow, giving the substantial volume of 85 million cubic meters annually.

Below Las Huertas, the Río Xiquila flows in a deep valley with steep slopes, hilly country, and some areas of bottom land; there are no permanent tributaries, and the drop to the Río Salado is 31 meters per kilometer. A large masonry aqueduct or canal was constructed during the late Palo Blanco (A.D. 400) and was in operation until the Conquest. This 6.2-kilometer-long structure, of which only scattered segments remain, consisted of masonry retaining walls reaching heights of 3 meters and of terraces cut into the steep slopes. The canal runs roughly parallel to the river's north side, and there is some evidence of a second canal on the other side. The canal was 100 centimeters wide and about 50 centimeters deep and originated, most probably, in a section of the valley just below Las Huertas, where a diversion dam, partially blocking the river, must have been located. As a testimony to prehistoric engineering, this is the place where the Papaloapan Commission, after much study, has proposed a dam for hydroelectric generation and irrigation water. Furthermore, the grade of this prehistoric aqueduct is 1.3–3.2 percent, well within the ideal parameters set by modern hydrologists.

The estimated irrigated area along the canal route was about 250 hectares, with another 250 hectares if terracing was used. The Papaloapan Commission has estimated that the Río Xiquila has an irrigation capacity of 3,000 to 3,500 hectares, more than is currently farmed using gravity water distribution along both the Xiquila and the Salado. Therefore, in this case, land was a limiting factor, and the 500 hectares must have produced two to three crops per year to make the aqueduct worthwhile (Woodbury and Neely 1972: 102–104).

Woodbury and Neely estimate the labor for construction to have been 8,580 worker-days. Ten workers could therefore have constructed the canal in 858 days, or two and a half to three years. The village sites along the canal could easily have provided the labor and may quite possibly have been within the borders of a local hierarchy such as a city-state or kingdom. A similar canal associated with the *cacicazgo* of Cuicatec supports this conclusion (Flannery 1983:336).

The aqueduct demonstrates that it was possible to plan and build a complex engineering structure to increase the agricultural production essential to maintain a nonagricultural elite supported by tribute. It is probable that the aqueduct functioned in a hierarchical economic system under centralized political control, as may have been the overall pattern in the entire Tehuacán Valley. The Río Xiquila was the only place where this type of construction could have been possible; the other rivers are too intermittent (Woodbury and Neely 1972: 112–113).

Agricultural Terraces and Borders

The use of terraces and borders to control runoff, as a means of maintaining and increasing soil moisture, was developed in early Palo Blanco times and continues to the present. These field systems were built primarily in the uneven terrain along the edges of the Tehuacán Valley and were most probably a response to the need to increase food production beyond the yields from the irrigated fields on the level valley floor. Two variations of the same technology have been identified: small valley systems consisting of low walls or lines of boulders across intermittent drainages, and hillside systems with horizontal walls along hilly flanks often overlooking intermittent streams. Both variants were constructed to control water flow from rain by slowing water to increase soil penetration, and also to reduce the soil erosion caused by reduction in natural vegetation from farming. Because both hill and valley techniques are still in use, as recognition of their effectiveness, it has been difficult to determine when they were first constructed (Woodbury and Neely 1972:113).

A typical hillside system, located two kilometers south of Texcala on the road to Zapotitlán Salinas, has terrace walls separated by 20 to 30 meters; each terrace is horizontal, perpendicular to drainage flow, and curved to follow the contour of the slopes. Lengths of 100 meters are common, and the average size of a field is 7.3 hectares, twice the size of a valley field. Sherds in association with traces of field houses in twenty-seven hill sites have been dated to the early Palo Blanco, although valley systems did not appear until late Palo Blanco. By

Venta Salada times the incidence of both was about equal (Woodbury and Neely 1972:113–114).

One of the best-preserved valley sites, also on the west side of Tehuacán near Texcala, is located near the head of a small canyon. The linear borders are 8 to 30 meters long at right angles to the channel draining the slopes. Sixteen well-defined parallel stone terraces 30 to 170 centimeters high, built of unmodified cobbles and boulders and cemented with an adobelike mortar, cover an area of 3 hectares. An aerial photographic survey has detailed forty-two fields with areas ranging from 0.2 to 4.0 hectares and an average area of 3.0 hectares. The differences in average area between hillside and valley variants can be interpreted as a response to the relative scarcity of land with adequate and manageable water supplies. At first the available open hillsides were used and then the smaller valleys and canyons with intermittent streams (Woodbury and Neely 1972:116–117).

Irrigation farming along the permanent streams and major springs was the primary form of cultivation, continued through Palo Blanco and Venta Salada times. The stone-bordered fields and terraces of the small hill and valley systems were an additional source of food and must have been relatively important in view of increasing populations and demand for agricultural produce. An examination of relative yields demonstrates the importance of increased areas for cultivation. According to Gibson, at the time of the Conquest a "normal" yield of corn in the Valley of Mexico, without using irrigation, was about eleven *fanegas*, or approximately one-thousand kilograms per hectare. An average family has been estimated to consume ten to twenty *fanegas* per year, which means that one to two hectares is sufficient for domestic needs (1964:309–311). Consequently, both valley and hillside systems were capable of producing a surplus. Each field system was probably a self-contained unit worked by a single farmer or family group. Furthermore, it was not necessary to construct the complete system before farming could meet domestic needs. Such systems were probably not as labor-intensive as had been previously believed (Woodbury and Neely 1972:125–126).

Large-Scale Canal Systems

A large number of canals no longer in use are in evidence on the valley floor. These extend from just northwest of Tehuacán to near the archaeological site of Venta Salada, some thirty-five to forty kilometers to the southeast. The term "fossil" canal was coined to describe their preservation by the extensive calcium carbonate deposition on the inner surfaces of the canal walls. Due to soil erosion, many of the canals now stand above the land surface. The canals spread out from three major springs: San Lorenzo, the side-by-side springs of Texcala (or La Taza) and Cozahuatl, and the now-dry Atzompa. Close to the springs some of the canals follow the paths of maximum grade in fanlike arrangement. Many, however, do not follow natural gradients but form a pattern of distribution to cultivated areas. The pattern of the fossil canals is similar to and often paralleled by modern canals in current use. The conclusion reached by Woodbury and Neely is that the fossil canals were part of an artificial and prehistoric water distribution network. It is quite possible that the first system was adapted and expanded from natural drainage channels. Numerous improvements have resulted in an extensive local knowledge of water control. The current canal system and irrigation technology are the end-product of a long sequence of innovations from simple origins (Woodbury and Neely 1972:127–130).

Although the main canals (1.5 to 3 meters wide and 0.3 to 1 meter deep) give the appearance of having been part of three separate systems, it is probable that the two systems emanating from the spring at San Lorenzo and the now-dry spring at San Marcos Necoxtla were actually one system 25 kilometers long, similar to the modern canal network from San Lorenzo. The connecting canals are missing because of intensive modern-day agriculture. Also, there is neither a flowing spring near Necoxtla nor any trace of springs gone dry. The idea of a complete system is more plausible in view of the need to end a system by providing a drainage for overflow. The modern canals end in natural drainage areas in the Ríos Zapotitlán and Salado. There is no drainage area between San Lorenzo and Necoxtla. It is also possible that all three systems could have been linked to form a single system 35 kilometers long (Woodbury and Neely 1972:130–132).

All the fossil canals have a spring as a source, and none are found on the east side of the valley, where a large amount of water comes in runoff. The flow from the springs is substantial, 300 to 500 lps, with less flow fluctuation than surface streams. The grade of the canals, well within modern engineering parameters for controllable flow, is 1.35 to 1.8 percent. The canals supplied relatively level fields that required only minimal borders and terracing. However, surface saline deposits must have been a problem in the past, as they are today (Woodbury and Neely 1972:135).

Because age cannot be determined by travertine deposits, clear indications of the dating of the canals have been difficult to find. Prehistoric construction is not in contention, and the best indicators of age are the proximity of salt production and dwelling sites. Numerous such sites are located near Necoxtla, where the pottery remains date back to late Palo Blanco through Venta Salada and into post-Conquest times. This chronology is supported by a few sherds found embedded in the canal structures (Woodbury and Neely 1972:135–139).

PREHISTORIC FARMING IN THE TEHUACÁN VALLEY

Of central importance to changing subsistence and settlement patterns are the techniques for irrigation agriculture and the sociopolitical nature of water management. The botanical evidence of the Coxcatlán Phase documents the first indications of small-scale farming. The moisture requirements of the cultigens made the deeper canyons and runoff bottom land ideal places for successful wet-season domestication. During the Abejas and Santa María phases, expanded cultivation is evidenced by settlement along waterways and canyon flanks.

The first identifiable farms and water control structures date about 700 B.C., during the first level of the Purrón Dam Complex. Previous to this time, small-scale irrigation must have developed along the tributaries to the Río Salado. The dam complex functioned for nine hundred years, until about A.D. 200. As far as can be ascertained, large earth-filled dams with a masonry veneer did *not* spread to other regions or continue in the Tehuacán Valley. Silting, flood damage, slumping from excess seepage, and problems of social control and management are probable reasons for abandonment.

The Purrón Dam did represent a tremendous development and expansion of irrigation during the Middle Formative[2] as was occurring in other parts of Mesoamerica. There was a corresponding population increase and an expanding proportion of agricultural products in the diet. The population increase from the Ajalpan to the Santa María Phase was from about 42 to 265 per square kilometer, and in the Palo Blanco it jumped to 1,100 per square kilometer with the total population estimated at 26,000. Field borders and terraces were a further contribution to the inventory of techniques used to increase the efficient use of soil and water resources.

Irrigation became more important for the expansion of food production, but reliance shifted to the use of small dams and diversion structures. Technical improvements were made in transporting water from perennial sources over increasingly longer distances using carefully graded canals. This was in contrast to storing large quantities of water from seasonal runoff. Mesoamerican prehistoric irrigation, in general, was principally concerned with the redistribution of current sources rather than with collection and storage for future distribution.

The current population is still dependent on systems developed in prehistoric times. New crops have been introduced and the social and political organization has changed, but the principal concern continues to be the building, maintenance, and allocation of water resources. The next section examines the social organization at the time of the Conquest in an attempt to make a historical link with current organizations of irrigators.

ETHNOHISTORY: TEHUACÁN AND THE CUICATEC *CAÑADA*

The archaeological record clearly documents a pattern of technological innovation and increased use of irrigation agriculture. Along with changes in technology, the social and political organization of these irrigation-based societies became increasingly complex, as evidenced by settlement patterns, architecture, tools, and pottery. By the time of the Spanish Conquest, the Tehuacán Valley was composed of at least three localized states with a dependent hinterland, and many questions come to mind when speculating about the nature of the social organization of these polities. Clearly, it was stratified—as evidenced by elite versus commoner housing—but what was the nature of the class structure and politics in relation to the control of productive resources of land, water, markets, and trade? How did this class structure originate and how was it maintained over time? What is the relationship between the pre-Conquest class structure and conflict over land and water during the colonial and recent periods? In this section we will attempt to answer these questions and examine numerous implications.

The best published account and analysis of land, water, and social organization is the late Eva Hunt's "Irrigation and the Socio-Political Organization of Cuicatec Cacicazgos" (1972). Using published and unpublished colonial sources and codices, she reconstructed the ethnohistory in great detail and painted a vivid picture of the sociopolitical organization of the Cuicatec up to the Spanish Conquest. Her principal sources were the *Relaciones geográficas* of the sixteenth century in the *Papeles de la Nueva España* (Paso y Troncoso 1905), the *Epistolario de la Nueva España* (Paso y Troncoso 1939–1942), and an old manuscript obtained in a Cuicatec village, which refers to periods before the Aztec conquest of the Cuicatec in A.D. 1510. The data and analyses are of particular importance to events in the Tehuacán Valley, not only because of the close geographical proximity, but also because of similar sociopolitical organization in the two areas. The Cuicatec orientation, in terms of both trade and politics, was north to Tehuacán (E. Hunt 1972:164).

Modern Cuicatlán is located about 110 kilometers south of Tehuacán in the northeastern part of the state of Oaxaca. The Cuicatec areas are bordered to the north by the confluence of the Río Grande and Río Salado forming the Río Santo Domingo. The southern border is the Almoloyas mountain chain, which separates the Cuicatec from the Valley of Oaxaca. The present population, as in the pre-Hispanic past, lives in two distinct ecological zones: the semiarid canyon, or *cañada*, along the Río Grande, and the high, temperate slopes of the Llano Español to the east. This ecological contrast is very similar to the valley floor and piedmont on the edge of the Tehuacán Valley. The *cañada* has permanent rivers, in contrast to the springs in Tehuacán. The piedmont in both areas consists of small plateaus, *barrancas*, and narrow ravines cut by intermittent streams. The argument for cultural proximity and similar sociopolitical organization is therefore strengthened by a long history of similar environmental adaptations (E. Hunt 1972:166–167).

The pre-Hispanic settlements and their surrounding territories roughly approximate the current municipal divisions, and the old political centers of the kingdoms are the present *cabeceras* (head towns and administrative centers) of *municipios*. Of forty Cuicatec settlements existing in the early colonial period, thirty-five have been identified in the original locations with similar names. These settlement patterns appear to have been determined by the location of water sources, thus forming a link with social organization and the management of productive resources; water control had and continues to have an important influence on the structural relations of the local population and its relations with other areas (E. Hunt 1972:169).

Irrigation is prehistoric in origin, and although detailed archaeological data and interpretations are lacking, the agricultural practices in the *cañada* were completely dependent on irrigation. Even though political affiliations have changed over the centuries, political power was and continues to be based on intensive irrigation agriculture. In terms of social evolution, it is difficult to assign primacy to irrigation as an independent variable, but in view of the fact that water control is crucial to any agriculture in the Cuicatec lands, competition over access to water and the protection of physical facilities must have been just as important in the past as they are in the present. It is the social, political, economic, and religious organization of these early Cuicatec irrigators—and, by inference, those of Tehuacán—that we now examine.

For an unknown period prior to the Aztec conquest in A.D. 1510 and the Spanish Conquest about ten years later, the Cuicatec region consisted of loosely centralized states called *cacicazgos*. The rulers were a distinct class of elites, a kind of nascent aristocracy, whose control over water was a source of political power. The local irrigation systems were not large, but their control, maintenance, and operation required personnel with technical and managerial skills. Centralization of decision making was a means to avoid conflict. Technically skilled personnel were essential to prevent the physical collapse of the system and served social control functions to avoid internal conflict leading to disrupted production (E. Hunt 1972:193).

The earliest systems were located in small hamlets, such as the *barranca* hamlets in Tehuacán and the Río Grande *cañada*, and were probably started as cooperative efforts by groups of domestic units. When the systems were very small, the technical and management activities were distributed among many individuals in the local populations. As the systems became larger, decentralized administration became less efficient and led to the development of specialized roles of authority over various irrigation activities. It is not possible to say when decentralization gave way to more specialized roles and eventual centralized control, but it is safe to say that in Tehuacán during the time between the first use of irrigation, prior to the Purrón Dam complex and the Conquest, there was an evolution toward more complex and specialized irrigation and political roles. By Palo Blanco and Venta Salada times, the centralized systems in the pre-Conquest Cuicatec and Tehuacán areas were probably analogous.

Land tenure prior to the Conquest did not consist of private property, and the Indian commoners, or *macehuales*, did not have permanent usufruct of parcels. Instead, each farmer was assigned a different plot of land every year. The local ruler, or cacique, was the land redistributor, and was assisted by a local commoner officer, the *tequilato*. Land under cacique control was a restricted patrimonial domain and not private property. The cacique did not own the land within his domain nor was he the absolute master over the *macehuales*. He was in charge of administrative and political tasks; in return he received his own parcel as a form of compensation. This type of communal tenure, a form of corporate control, has been preserved into modern times in the form of *tierras comunales*. The *macehuales* did voluntary work on the cacique's parcel and made no payments for land redistribution.

The cacique had a number of other privileges and tributes: personal servants, firewood, cloth, and exclusive right to eat deer meat, drink cacao, and wear fancy sandals. After the Conquest, he could use a sword and ride horses, privileges that were denied to the commoners by the early Spanish colonial administrators. A cacique lost his privileges if he moved away from his administrative domain (*estancia*), and it appears that residency was a requirement for this type of political control throughout Mesoamerica. Also, the cacique could not rent *estancia* land to outsiders, but he did have authority to sell water (E. Hunt 1972:200).

The annual land redistribution prevented the same commoner families from monopolizing the best parcels. This may have served to keep the wealth of the *macehuales* at a homogeneously low level and prevented the establishment of wealthy commoner families who could have threatened cacique power. Caciques and communities of *macehuales* were in continuous corporate control of *estancia* lands, which amounted to a form of insurance against internal factionalism and land fission. Of course, this also served to increase the power of a minority elite class with administrative control over communal lands and virtually exclusive rights to irrigation water.

Cacique political power was also derived from the control of military activities and a monopoly of temple services. The cacique appointed military officers and temple priests, both of whom were recruited from aristocratic kinship lineages. Warfare was important in the maintenance and extension of control over both land and water.

Religious activities were also important, as evidenced by the large number of ruins of ceremonial complexes. The religious system legitimized the power and control of the cacique, and there was a direct relationship between political power and the economy of irrigation agriculture. The caciques did not accomplish this by controlling large-scale irrigation works but through military and religious leadership and the control of scarce resources—water, the redistribution of land, and scheduling and organization of community labor (E. Hunt 1972:226).

The Cuicatec cacique could quite possibly have been a provincial counterpart of the Aztec *calpullec*, who was in control of communal *altepetalli* or *calpulli* lands. The cacique's personal plots could have corresponded to the Aztec *tecpantlalli*, or office lands. Land tenure also extended to temples and religious schools that were assigned agricultural fields called *teopantlalli* (E. Hunt 1972:227).

Alfonso Caso (1963) asserts that it is not valid to project social organization and land tenure forms from Tenochtitlán to the tribute-paying provinces, because of lower levels of social differentiation. But clearly, the head towns and larger villages in the Tehuacán Valley and the Cuicatec had more complex social organization than mountain communities, albeit less than existed in the center of the emerging Aztec state. The Cuicatec *cañada* and Tehuacán kingdom had fewer social strata and a smaller wealth differential between the elite and commoners (E. Hunt 1972:233).

The elite were organized into ambilateral descent groups divided into ranked minor ramages.[3] There were three levels within the localized aristocracies: the royal house descent line from which caciques were chosen, junior and cadet branches that ruled over commoner subject settlements, and collateral lines without succession privileges to cacique and overlord positions but intermarried with the elite. This last group held the lower political, administrative, and military positions. The elite were clearly rulers with internal stratification and were, as a group, above the descent groups of the commoners. Although these elite ramages were class endogamous, they did intermarry with elites in other ethnic groups and territorial divisions. Local caciques arranged marriages very much like eighteenth-century European royalty. Using Paul Kirchhoff's (1952) terminology, these descent groups were part of a conical ambilateral clan system suggesting a modified *calpul*-type organization (E. Hunt 1972:238).

The lower class, or commoners, also had three differentiated ranks. In the first, and by far the largest, group were the free commoners, who in all probability resided permanently in the corporate village community. There are no data on kinship organization or on whether they were village endogamous or exogamous (E. Hunt 1972). We do believe, however, that the commoner descent groups were also ranked according to differential access to land and water. Even though land was redistributed annually, we find it hard to believe that this resulted in constant equity over time. Furthermore, and possibly even more important, there are no data on how the caciques distributed water, a crucial variable for differential agricultural output in the Tehuacán Valley and the Cuicatec *cañada.*

The second commoner rank was a small group comprising the *tequilato* administrators, who were definitely of commoner origin, but who represented a link between the two macro social classes. The third group was a class of slaves and prisoners of war, who were lower than the free commoners. This group was probably a source of sacrificial victims.

Within the commoner class as a whole there were craft specialists, but there is no direct evidence on the relative rank of the specialties or on rank within the common class in general. One can safely conclude that the descent organization and marriage prescriptions resulted in two separate macro-classes. Internal subdivision has been demonstrated for the elite and was most probably also the case for the commoners. The local territorial divisions supported a hierarchy of unequal social positions using descent affiliation. The relative ranking of the descent groups and territoriality maintained both vertical and horizontal division and established links with other territorial and ethnic groups. This analysis may imply that the Tehuacán and Cuicatec political elite had been mixed by intermarriage with aristocratic groups from areas of higher civilization, such as the Valley of Mexico and Puebla-Cholula. A further possibility is that provincial areas, with less complex techno-economic and political organization, evolved state-level organization by virtue of outside influences (E. Hunt 1972:205).

Larger states were not formed in the Cuicatec and Tehuacán because tendencies toward fusion into larger units must have been counterbalanced by competition and conflict over water. The ethnohistorical sources show that such rivalries resulted in fission along sibling divisions in ruling descent groups. Consequently, fission reestablished the population balance between land and agricultural resources. After the Aztec conquest, war was forbidden and local territorial divisions became permanently established. Caciques became officially designated by the state, leading to longer periods of stability. Some caciques obtained new rights and privileges to ensure their loyalty to the Aztec state. Later, the Spanish continued to support local rulers, thus further preventing revolts and violent conflicts over resources. The Conquest resulted in the termination of the natural and relatively independent evolution of the Cuicatec and Tehuacán political systems (E. Hunt 1972:237).

The power held by the elite was derived from, and supported and maintained by, the Aztecs and later by the Spanish. The Aztec policy was indirect rule, which was continued by the Spanish until the local caciques could be replaced by Spanish administrators. By the end of the seventeenth century the Indian aristocracy had lost its importance and its power base was destroyed (E. Hunt 1972:242).

The decline of the native upper class as it was replaced by Spanish civil servants, clergy, and military authorities did not eliminate all local power, but rather left a degree of autonomy for the Cuicatec Indian commoners. Similar trends have been documented for Tehuacán, Puebla, and Morelos. Since colonial times, the native populations have maintained some degree of linguistic and cultural identity, even though their settlements, villages, and towns have mestizo populations. Competition over access to land and water continued, and the resultant establishment of new caciques quickly replaced the traditional nobility. Modern-day caciques are not very different in that potential adversaries are still united through intermarriage, and the new elite controls local PRI politics, land and water administration, and commerce and ties local areas to regional and national centers of political and economic power.

The evidence from Cuicatec, Tehuacán, and Teotitlán del Camino shows a long history of elite involvement in irrigation management. Some investigators have assigned to irrigation the role of independent variable—the sufficient cause for the development of complex societies at the time of the Conquest. As stated above, it would be simplistic to assign sufficient cause to a single variable. Irrigation, no doubt, was responsible for increases in agricultural productivity, the domestication of new plant varieties, and eventual population increase, but was not the sole cause for civilization, political integration, and centralization. A more probable and sound conclusion is that irrigation was a necessary rather than a sufficient cause for the development of civilization and pre-Hispanic city-states in the highlands of Mexico (E. Hunt 1972:244–245).

4.

POST-CONQUEST CONFLICT OVER LAND AND WATER

WRITTEN IN COLLABORATION WITH LUIS EMILIO HENAO

FROM CONQUEST TO REVOLUTION

THE POST-CONQUEST history of the Tehuacán Valley is filled with the sacrifice and sweat of the Indians who inhabited the rich valley.[1] As in many other productive regions of the Spanish colonial empire, the conquerors set up a system of conscripted labor and services based on the *encomienda* and the *repartimiento*. The *encomienda* was an "official consignment of groups of Indians to privileged Spanish colonists" (Gibson 1964:58), which allowed the Spaniards to extract tribute or tax from Indian populations within their domain. The *encomendero* (landholder) was supposed to provide protection for the Indians as a trustee and to convert them to Christianity.

The tax system was often abused by the Spaniards. The *repartimiento* was a mechanism created to ensure labor for the haciendas by forcing Indians to work. Although they were supposed to be paid a minimal wage, the system was thoroughly exploitative, effectively breaking up subsistence production in many regions of the empire by forcing Indians to leave their fields at critical points in the agricultural cycle. The results were famine, disease, and drastic changes in traditional forms of social organization.

In this section we sketch the colonial history of the valley, the change resulting from the creation of the haciendas, and the impact of the systems imposed to extract taxes and labor from the indigenous population. There are three major points we wish to make. First, from the very beginning of the colonial period, the state played a dominant role in setting the conditions for economic, political, and religious organization in the valley communities. Although some of the social science literature has treated communities as if they functioned as isolates, this has seldom been the case in Mexico. Second, the struggle for the basic resources in the Tehuacán Valley was as desperate as in many other areas of Mexico. But what is striking about the Tehuacán Valley is the persistence and resilience of the Indian population in its struggle to regain lost water and land rights from the haciendas. The battles were carried out in the courts, through physical confrontation, and through the sacrifice of labor in the fields. From the colonial period to the Revolution, despite the brutal and taxing system of the Spaniards and later the mestizos, the Indian communities repeatedly regained access to water and land resources by means of cooperative associations. Ultimately, after the agrarian reform, the Indians were once again a major force in agriculture.

The persistent efforts of the Indian population to gain control of water and land were periodically marked by success. This was, of course, a qualified success, because only a segment of the Indian population benefited, and often success ensured mere subsistence. Ironically, success even strengthened the dominant forms of exploitation and control. Particularly after the liberal reforms, the growing stratification within communities fragmented Indian efforts to respond as a larger communal or ethnic group to the mechanisms of domination.

Independence did not alter the basic nature of the conflict in Mexico or the Tehuacán Valley. As the demand for exports such as sugar grew and the domestic market for grains expanded, hacienda owners pressed to extend their land and water holdings at the expense of the Indian communities. This process reached its zenith during the Porfiriato, in the late nineteenth and early twentieth centuries.

The development of the hacienda system in Mexico did not come without opposition. The struggle between haciendas and the Indian communities in the Tehuacán Valley was played out in a variety of forms ranging from overt violence to sullen resistance.[2] Again, one is struck by the ability of the indigenous valley population to cope with the intrusion of the haciendas and in some cases even to regain land and water rights. The nature of the system forged by conflicting forces left an important legacy for the present. The indigenous population's control over productive resources and the contemporary forms of organization have their roots in the struggles of the past.

The third major point of this section is that the forces of domination contributed to stratification within the Indian communities based on differential control of basic resources. While communally owned land and water did play a major role in village life until the reforms under Porfirio Díaz, the Spanish colonial structure imposed increased stratification. The reforms of the Porfiriato and the struggles for power during and after the Mexican Revolution led to greater stratification, when viewed in the context of the expanding national dependent capitalist economy. Despite these pressures, community and ethnic identity have been maintained.

The Spanish easily took control of the Tehuacán Valley as the major indigenous leaders capitulated without battle. In 1540 much of the region was incorporated into the *encomienda* of Antonio Ruiz and the remainder was controlled by the viceroy (Veerman n.d.:36). The Spanish made Tepeaca the regional capital until 1660, at which time the city of Tehuacán, which had evolved as a trade center, was designated the political capital.

The Spanish occupation of the valley brought many profound changes as Spaniards quickly usurped the major land, water, and labor resources. The occupation transformed the valley both spatially and structurally. A series of forced relocation projects, culminating in a major relocation of the indigenous population in 1602, reorganized many communities. In some cases, entire villages were relocated; in other cases, smaller villages were transplanted and merged with other communities.

In 1622, after the demise of the large *encomiendas,* the hacienda system was created, forged from the land and water of the Indian communities. The haciendas quickly converted the land for the production of wheat and livestock. Corn, beans, and squash—traditional subsistence crops of the Indians—continued to be the base of the Indian agricultural system maintained on marginal lands.

To help administer the valley, the Spanish brought in indigenous leaders from the state of Tlaxcala. By 1725 more than 130 caciques were residing in the valley (Paredes Colín 1953:96). In exchange for their services, the caciques were given certain privileges, such as land they could sell or leave to their descendants and exemptions from paying taxes. Because the caciques spoke Spanish, they served as brokers between the Indian population and the colonists. Over time, the caciques were able to accumulate considerable wealth and independent power, distinguishing them from most other indigenous groups in the valley.

The development of the hacienda system in the valley led to an endless series of conflicts between the indigenous communities and the Spanish over land, water, and labor. Although the mechanisms for gaining control of these resources varied, the indigenous population struggled against each threat to its resources. Forced labor, a plethora of taxes, and confiscation of land and water forced some families to flee the valley for the sierra. Other families were forced to leave their communities and become full-time employees on the haciendas, giving up their rights to communal land and water. Nevertheless, many families managed to remain in the villages, with members working periodically for the Spaniards, yet producing enough surplus on their agricultural plots to pay taxes and, at the same time, participating in the village religious and political life. By maintaining the viability of community organization, they were able to fight the expansion of the haciendas' efforts to take land and water. Weak, and often disregarded, documents such as codices, land titles, or evidence of early community occupation of lands were the only legal protection communities had from the expanding haciendas.

Spanish administrators were granted large tracts of free land throughout the sixteenth and early seventeenth centuries. By the mid-seventeenth century the indigenous towns of Altepexi, Ajalpan, Chilac, Zinacatepec, Miahuatlán, and Coxcatlán were surrounded by haciendas that had acquired land and water that had previously belonged to these communities. The haciendas were established in locations blessed with access to irrigation water and canal systems previously used by the Indian communities. From the very beginning the haciendas provoked a series of lengthy conflicts over water: rights of use and possession, water thefts, measurement alterations, canal appropriation, and canal destruction. Efforts by the colonial government to legally arbitrate land titles were opposed by many hacienda owners, who often illegally controlled land and water.

By the end of the seventeenth century there were seventeen haciendas in the southeast part of the Tehuacán Valley, and the major indigenous communities continued to have their land and water systematically confiscated. San Gabriel Chilac lost land to the haciendas of San Francisco and San Gabriel. The Santa Cruz and San Miguel haciendas took water from the spring Cozahuatl and land from the communities of Chilac, Altepexi, and Ajalpan. Water from the spring Atzompa was taken by Hacienda Calipan, as were lands from San José Miahuatlán. Hacienda San Lucas Venta Negra captured land and water from San Sebastián Zinacatepec.

In the southern end of the valley the early cultivation of sugarcane accelerated the process of transformation. A sugar *trapiche* (cane juice extraction mill) was constructed in Calipan as early as 1703. The hacienda expanded after 1760 when a new owner forcibly pushed Indians off their land. Another sugar mill was established at San José Buenavista, on the northern side of Ajalpan. Sugar was planted extensively on Hacienda San Lucas Venta Negra just west of the Indian community of San Sebastián Zinacatepec. Permission from the colonial government to create a sugar mill was strictly contingent on not using Indians to do skilled work. Nevertheless, Indian labor was increasingly used in the mills because of the scarcity of other labor in the area.

The scarcity of free labor was a constant problem for hacienda owners, who pressured the government for permission to obtain workers from the indigenous communities. In 1708 the government dictated a law that ordered the natives to work on the haciendas for a minimal salary. This was met with strong protest from the communities, which systematically refused to deliver workers to the haciendas. Like the haciendas, the communities pleaded that they needed labor to sustain their agricultural production, which provided a large part of the Tehuacán region with vegetables. Furthermore, the workers were also needed for the cultivation of communal lands in order to pay the taxes. Hacienda owners protested that without indigenous labor they could not maintain production of sugar and wheat. Finally, in 1739 the colonial government decreed that each community must send a quota of workers to the haciendas. Again, the indigenous communities refused to participate, leading to increased conflict between the indigenous population and the Spanish. In 1759 another decree demanded that Indians work on the haciendas; if ill, they were to send substitutes. This decree was enforced by a prison sentence for those Indians who refused to yield. Although many Indians continued to refuse to work on the haciendas, the decree made it possible for the Spanish to expand their control over the irrigation system. When hacienda owners took more water from the Indians, the Indians were reluctant to protest in Tehuacán, fearing they would be imprisoned for refusing to work on the same haciendas that were taking their water.

Working conditions on the haciendas were grueling and often brutal. Indians were required to work for extremely low wages from sunrise to sunset under the strict supervision of the caciques. Transgressions in the sugar mills, factories, or haciendas could result in the offenders' being sold as slaves. There are reports of Indians fleeing the valley to avoid working on the haciendas.

Meanwhile, the colonial government was restructuring the indigenous communities and the ties between them. Throughout the eighteenth century, and climaxing in 1764 by decree, Indians were prohibited from migrating or traveling to other towns without administrative permission. The decree further strengthened the Spaniards' ability to control the population and allocate labor.

The taxes demanded of the Indians by the colonial government included a percentage of agricultural production as well as a regular household tax. But not all Indians, particularly leaders, were included. In addition, community civil and religious leaders were exempt from mandatory labor requirements on the haciendas while they held office. They were also given an extra quantity of water, called "compensation water," as reimbursement for the time spent in service to the community.

By the late eighteenth century, a small group of mestizos, usually Indian-Spanish mixed-bloods, had come to live in the communities. But at no time did the Indians consider them to be part of the community. They were not required to pay taxes or work on the haciendas and were actively engaged in commercializing crops raised by the Indians. The mestizos, also being merchants, took over the principal trades such as metalworking and tanning.

In addition to other demands, the Indians in the valley were required to work on construction projects of public works as well as religious buildings. The obligations of *faenas de casas reales* date back to the sixteenth century (Paredes Colín 1953:71), and were in full force at the end of the eighteenth century. Major construction projects in Tehuacán were divided between the communities, which were expected to provide materials and labor.

The plight of the Indians did not stop with taxes and forced labor. Periodic contributions were extracted from the communities for the maintenance of the army, the sustenance of the San Lázaro Hospital in Tehuacán, and special expenses for church or government holiday celebrations. This tax was taken in various forms, including forced participation in the lottery. In addition to the government taxes, the church siphoned off what little surplus remained in the communities. Masses, particularly burial masses, were very expensive. When families were unable to pay these expenses, land, water rights, and even homes were confiscated by the church.

The tremendous pressure on the communities and subsequently on Indian families for labor and tax money brought about a series of changes. More land was brought into production, requiring more hours of work from all members of the family. Whereas formerly agriculture had been based primarily on irrigation, more land was now planted with the hope that the crops would receive enough moisture during the rainy season to produce. Unfortunately, the rains were often capricious and lacking.

Despite increased labor, community agricultural production declined as the communal water was lost to the haciendas. The towns tried to acquire water by renting it back from the owners of the haciendas. Although the effort was seldom successful, in some cases agreements were reached. For example, in 1768 the town of Ajalpan was able to rent water from cacique Jacinto José Espinoza of Tehuacán, who controlled an important spring on his hacienda located in San

Marcos Necoxtla. Of course, this water had been used by neighboring communities before the Spanish arrived. Acquisition of this water made it possible to plant vegetables, which were sold throughout the province and as far away as Veracruz. Efforts by Ajalpan residents to rent water from the Hacienda La Huerta during the same period were unsuccessful.

Despite the pressure from the Spanish, the Indian communities maintained their sociopolitical integrity and were able to regain some of the land and water lost to the Spanish. In 1765 the towns of San José Miahuatlán and San Sebastián Zinacatepec were able to buy land previously held as an hacienda. The hacienda was located between the two towns. In 1709, San Gabriel Chilac bought the Hacienda San Miguel from the Marqués de Buenavista (Henao 1980:71). These purchases strained the communities' financial resources, but were critical in allowing them to expand agricultural production.

The Indians' struggle for survival was further complicated by a series of epidemics of European diseases. In 1776 one of the most severe plagues killed 13 percent of the economically active population in Ajalpan and almost equal percentages in the other major communities. Many of the survivors were immobilized for the period of their illness, leaving the labor force depleted. Yet despite the epidemics and the grueling work conditions, the towns of the valley maintained relatively large populations. In 1791 Ajalpan had 2,181 Indian residents and 420 Spaniards, mestizos, and mulattoes; Zinacatepec had 1,948 Indians; San José Miahuatlán had 2,166 Indians; and Chilac 1,576 Indians (Veerman n.d.:41).

At the beginning of the nineteenth century, the haciendas in the Tehuacán Valley were based primarily on the cultivation of wheat and corn, as well as on the raising of goats for annual slaughter. Though they varied in size, no single hacienda dominated the valley. This changed in the middle of the nineteenth century with the expansion of the Haciendas San José Buenavista and San Francisco Javier Calipam (later Calipan). Both haciendas annexed water and land resources from their neighbors as their owners strove to produce sugar for the growing domestic and international markets.

Hacienda Buenavista, located near the town of Ajalpan, was transformed after 1848 when a new owner installed a rum factory and expanded the sugar mill. The owner rapidly increased the cultivation of sugarcane, buying other haciendas such as Zavaleta and Xooxtipa, as well as leasing land from other haciendas and Azalpan communal land. Hacienda Buenavista continued to expand until it controlled 38,625 hectares.

The second major sugar hacienda was Calipam. The owner wrestled the critical water resources of the spring La Ciénaga from indigenous farmers who had previously purchased the water rights from the Haciendas San Pedro and Nopala. Soon afterwards, the hacienda was sold to General Mucio P. Martínez, governor of the state of Puebla. This occurred during the Porfiriato, and the governor had free rein to harass villagers to increase his fortune. Families from San Sebastián Zinacatepec and Coxcatlán were the primary source of labor for the hacienda during this period. People were forced to work from dawn until dark and were not allowed to talk with one another while they worked.

As sugar production expanded, the demand for seasonal labor for the harvest also grew. Although wages were minimal, Indians were forced to pay taxes. Those without land and water were unable to cultivate cash crops. The only alternative to jail was seasonal labor. In addition, the vagrancy laws practically allowed haciendas to capture workers. By 1880, Hacienda Buenavista had some three-hundred seasonal workers during the sugar harvest, the *tiempo de zafra,* from February to May. The workers came primarily from the agricultural communities of Ajalpan, San José Miahuatlán, Altepexi, and Chilac. The wages were twenty-five centavos per day, and remained thus for fifty years, when they were raised to thirty-one centavos.

The growth of the haciendas was facilitated by national legislation decreed by the government of Porfirio Díaz from 1880 to 1911. This period witnessed profound and widespread transformations throughout Mexico. The legal structure for these changes was grounded in three major laws. The first of these was the Ley Lerdo, passed in 1856, which prohibited corporations, civil or religious, from owning property not necessary for the function of the organization. Once written into the constitution, this law not only made it possible for villages to divide up communal lands among their members, but forced many to do so. In the early stages, villages having communal lands at the time of the Conquest were exempt from the law. But the law was expanded in 1889 and 1890 to include all villages. The social impact of this legislation was devastating as owners of haciendas and land speculators were able to either buy or force Indians into selling or transferring land titles.

The second major law was the Idle Lands Law (Ley de Tierras Ociosas) which stated that idle land could be confiscated. The intent of the legislation was expanded in 1894 to make it possible to seize any land without a title. The dominant class rapidly grasped the opportunity, restructuring the patterns of land tenure in the country.

"One estimate indicates that over two and one-quarter million acres of good land, representing the means of livelihood of tens of thousands of Indians, passed from indigenous communities to the haciendas. . . . By 1910, less than one percent of the families owned or controlled about 85 percent of the land" (Cumberland 1952:3–28, in Hanke 1974:476).

The third law, of critical importance in arid regions, gave the government control of water rights throughout all of Mexico. Government officials then reallocated water to their friends, especially those who were expanding production of crops such as sugarcane.

The impact of these laws in the valley was devastating. The communal land and water in the town of Altepexi were converted to individual holdings, many of which were sold to neighboring haciendas. Ajalpan sold part of the community water to the Hacienda Buenavista. Hacienda Calipam bought tracts of land formerly owned by Ajalpan.

The division of communal land met resistance from a segment of the Indian population, a protest that was finally settled with armed force by the government. The breakup of the communal holdings into individual holdings also made it possible for some of the more powerful Indians to accumulate land from their poorer neighbors.

The haciendas with abundant land and water resources developed new methods to generate income without making significant investment: renting out land and water resources, and sharecropping. The Hacienda San Lucas rented its water from the springs Atzompa and Tilapala, and its land to the people of San Sebastián Zinacatepec. Although this relationship dated back to 1641, it was greatly expanded by 1915, when the hacienda owners rented over 1,000 hectares of irrigated land and 1,700 hectares of rain-dependent land. The major crops were corn and sugarcane. Renting out their land allowed the absentee hacienda owners to avoid the risk associated with agricultural production and the conflicts with labor, yet still make money. Interestingly, this is the only hacienda in the valley that was destroyed and taken over during the Mexican Revolution.

Sharecropping also proved an important method for tapping Indian labor. Families were given permission to grow crops on hacienda land in exchange for half of the corn raised on the land, resulting in decreased cost and risk for the landholders. Sharecropping increased within the communities as a result of the Reform Laws and the discrepancies in land and water ownership. There was even sharecropping between Indians of different communities. By the beginning of the twentieth century, the vast majority of the Indian communities in the Tehuacán Valley were composed of two major groups—peons and sharecroppers. A small but significant number of families in the major towns had emerged that controlled greater amounts of land and water than they had before the liberal reforms (Henao 1980).

Yet even during this period there is evidence that Indians were pooling their resources to regain land. In 1875 residents of San Sebastián Zinacatepec formed a society to purchase the land of the Haciendas Nopala and San Pedro. The same society bought the water rights from the haciendas that controlled the spring La Ciénaga. It is not certain how the Indians generated the capital to make the purchases, but the money probably came from two sources—wage labor and the commercialization of corn.

The formation of the association through which individuals could gain special access to resources was an important step. It transformed differences in resource control based on derivative power from the Spaniards or mestizo *hacendados* into independent power based on privately owned means of production.

During the last half of the nineteenth and the beginning of the twentieth centuries, the haciendas and the indigenous groups began an intensive effort to increase water resources; this appears to be the first time that chain wells were constructed in the valley. The opening of the chain wells, many of which eventually were controlled by revolutionary generals, made capital accumulation possible for a number of families. Numerous groups of sharecroppers and peons banded together to contribute labor or money to finance the construction of the chain wells. They were organized as water associations, *sociedades explotadoras y distribuidoras de aguas*, or societies for water use and distribution for irrigation. Many of the early chain wells failed, but those built by associations in Chilac, Altepexi, Zinacatepec, and, to a lesser degree, in Ajalpan were successful in developing important resources. This water was critical to the expansion of commercial farming in the valley.

During this same period, a farmers' cooperative was organized in San Sebastián Zinacatepec. The cooperative rented water from the owner of the Hacienda San Andrés, located near the city of Tehuacán. As the water flowed down the valley, it was used to generate electricity for Tehuacán and for the Xaltepec thread and textile factory in Altepexi as well for irrigation by private individuals in Chilac, Altepexi, and Ajalpan.

THE MEXICAN REVOLUTION

The period of the Mexican Revolution was one of great agitation but relatively little military action in the valley. Life on the large haciendas was not significantly altered, although there was constant pressure from villagers and workers for land. It was the villages that experienced the burden of violence. Of the people we interviewed who were alive during the Revolution, many felt that they were victims of both Zapatistas and federalists. As one old man said, "Both sides only came to commit outrages, to plunder and to rob."

The revolutionary movement in the valley was instigated and directed by a group of owners of medium-sized farms and ranches in the valley and the sierra. Uprisings occurred in 1910 and lasted through 1916. Two major revolutionary parties existed, the first headed by General Francisco Barbosa of Ajalpan and the second by General Prisciliano Ruiz. The Barbosa faction favored the return of water and land to the communities while the other faction advocated the government position (Henao 1980:128). The two groups battled each other for control of the valley.

After 1915 there were various attempts to invade haciendas, but either these were repelled or the occupation was short-lived. One of the most successful occupations was carried out by residents of the community Nativitas, who took possession of land and water belonging to Hacienda Xochitlapan. In 1915, the residents of Altepexi were given water both from the spring La Taza, which had been controlled by Hacienda Buenavista, and from the spring Cozahuatl, which had previously been owned by Hacienda San Francisco. The recapture of the water and land did not last, since the revolutionary government was more interested in reestablishing the status quo than in redistribution. There was a significant reshuffling of power as the valley communities became politically independent of the city of Tehuacán and the revolutionary leaders emerged as major land and water holders in the valley.

The Revolution strengthened the economic and political position of the local bourgeoisie who had been revolutionary leaders. As early as 1917 General Alfredo G. Machuca bought Hacienda La Trinidad. This created a conflict with the campesinos of the community of Nativitas, who had claimed the land and water. Both parties petitioned the president for ownership of the lands. The official decision returned the land to the hacienda and the general. A few years later the hacienda was sold to the revolutionary general from Coxcatlán, Donato Bravo Izquierdo. General Barbosa took control of Ajalpan's land for his troops, but was finally forced to return it to Hacienda Buenavista. He then bought the land adjacent to the village San Marcos Necoxtla and expanded corn cultivation. With the construction of a major chain well by his "volunteer" labor force of campesinos, he became a key water merchant in the valley. To this day his family is still the principal water owner in Ajalpan (Henao 1980:130).

The fighting during the Revolution did not destroy the hacienda system in the valley. Only two haciendas were burned—Xoncotipa in the sierra and San Lucas near Zinacatepec. The first was burned at the very beginning of the Revolution. San Lucas had been an unproductive hacienda since 1886, renting out land and water. It was destroyed in 1914 not by the revolutionaries, but by General Higinio Aguilar and government troops. The residents of Altepexi, San Sebastián, and Ajalpan thereupon took animals and furniture from the hacienda. Hacienda Buenavista did not lose its lands or water, but at times served as a sanctuary for the revolutionary forces of General Bravo Izquierdo. Haciendas San Francisco and Xochitlapan recovered their lands and water, and sugarcane again became the dominant crop on La Trinidad (Henao 1980:130). Major haciendas such as San Andrés and La Huerta retained both land and water. Production was never seriously interrupted at Calipam, the dominant sugar hacienda in the valley.

Although the consequences of the Mexican Revolution did not lead to the immediate takeover of the haciendas by the campesinos, another force was gaining power that led to new forms of labor organization in the valley and ultimately played a role in the confiscation of land and water from the haciendas. This was the development of the labor movement in the valley. The first effort to organize unions in the Tehuacán Valley occurred in 1899, when the owner of Hacienda San Francisco decided to establish a thread and textile factory, San Juan Nepomuceno Xaltepec, in the town of Altepexi. Many of the peons from his hacienda were moved to the factory, which by 1907 had over one hundred workers. The factory became the site of the first union organization in the valley. The Confederation of Mexican Workers (Confederación Regional de Obreros Mexicanos—CROM), established in 1918, organized two unions in the valley as part of its early mobilization. The first was at the Xaltepec factory and the second on the Hacienda Buenavista, then one of the major sugar-producing haciendas in the valley.

CROM had a more difficult time in organizing Hacienda Calipam, where the millworkers provided the nucleus of personnel dedicated to forming the union. The owners of the hacienda forbade all types of meetings and went as far as to prohibit groups of men from talking, even about work on the hacienda or mill. The hacienda's security was turned over to the military sector, which posted a small detachment of men at the mill. Clandestine meetings were held in the fields at night, and with the help of CROM representatives from the Baker's Union in Orizaba the union was organized. The hacienda owners had management thugs beat up union leaders and dismissed them from the mill or hacienda for "subversive" activities. The men who were fired found work in the area and returned on weekends to help organize.

It was only with the alliance and support of General Barbosa of Ajalpan that the union was finally created. The CROM organizers from Veracruz and the union leaders of Calipam met with General Barbosa, whom they persuaded to provide protection for union leaders and to put an end to their persecution. Aided by growing regional support for the union and a more supportive national climate, the union was finally accepted by the hacienda owners. It quickly won a reduction of the work day from twelve to eight hours, a 25 percent pay increase, and medical services for the workers.

The coordination of the unions on the sugar *ingenios* of Buenavista, Calipam, and Central de Ayota, located just across the state line in Oaxaca, was an important step in the growth of the union movement. In all of these cases, the unions were built around the permanent workers and generally excluded the seasonal workers, who came for three to four months to cut the sugarcane. At Buenavista, seasonal workers tried to unionize, but the local permanent workers, leaders with support from the CROM hierarchy, blocked their efforts.

The permanent workers coordinated seasonal workers and identified with management interests. Despite this, temporary workers from San Gabriel Chilac who worked in the sugar harvest at Calipam formed a union. At first the permanent and temporary workers at Calipam coordinated their union activities. All of the unions were supported by CROM, but by the end of the 1920s CROM had become weak nationally, losing massive numbers of members and much of its state revenue (Cockcroft 1983:122).

When Vicente Lombardo Toledano formed the Confederación de Obreros y Campesinos de México (COCM) in 1933, to break with the Calles-controlled CROM, efforts were made to mobilize the seasonal workers in the Tehuacán Valley. The permanent workers' unions at valley haciendas remained committed to CROM, however. As the conflict between the two national unions intensified, the hostility served to augment already-existing antagonisms between permanent and temporary workers at the local level. The sugar mills remained in private ownership, as the government utilized them as a mechanism to balance and control the peasantry. The cane growers joined the Confederación Nacional Campesina (CNC), an arm of the government, but the seasonal workers who were hired to harvest the cane were never incorporated into the CNC.[3]

THE AGRARIAN REFORM: LEGAL CODES AND INSTITUTIONAL STRUCTURE

In order to understand the organization and operation of *galería* irrigation systems in Mexico, the structure of ownership, use, and control of land and water must be defined and examined in relation to the state. Once the relations of property and other resources have been established, the particular state agencies and their position in the state organization must be known in order to understand the institutions and activities that relate to irrigation agriculture at the local *municipio* level. To accomplish this, we describe the agrarian reform and the constitutional decrees that resulted from the Mexican Revolution and that form the basis for the present relations of landed property and access to water. We also examine the bureaucratic structure of the state, which links local agencies with the federal bureaucracy responsible for the administration of both land and water resources.

The present form of land and water ownership and the state institutions that administer these resources are the product of a chain of events that began with the Mexican Revolution and, more specifically, with the Decree of January 6, 1915 (Decreto del 6 de enero de 1915). The legal codes defining landownership and the use of natural resources have been modified since 1918, but the Decree of 1915 has provided the basis and direction for the subsequent changes.

The decree stated that all village lands that had been appropriated by illegal means prior to the Revolution would be returned to their rightful owners; villages with insufficient land and without prior claims to ownership would be granted enough land to meet their agricultural needs. The decree was meant to satisfy the needs of the landless rural population whose plight had become the theme of the Revolution. The agrarian problem was clearly defined as the need to redistribute land and resources by breaking up large landholdings in order to meet the demands of the rural populace. All subsequent decrees, constitutional articles, and legal codes define and limit the implementation of the agrarian reform as stated in the Decree of 1915.

The most important reform milestones resulting from the Decree of 1915 are Article 27 of the Constitution of 1917 and the Agrarian Code of 1934. Numerous subsequent presidential decrees have imposed restrictions or modified requirements, but the basic legal definition of private property, the ownership and use of natural resources, the distribution of land, and the organizational requirements for use and transfer of land and water have been defined and implemented since 1915. The legal codes define eligibility and conditions whereby private individuals and groups may have tenure of land and access to natural resources.

The holding of land by individuals and the exploitation of groundwater resources in *municipios* such as Altepexi, Chilac, and Ajalpan are direct results of the agrarian reform. The numerous changes have been primarily concerned with the exact hectarage that individuals can hold, from whom and how much land can be taken for redistribution, and the structure of the requisite administrative bureaucracy on the local, state, and national levels.

Large-scale land redistribution did not begin until 1920 with the Alvaro Obregón administration, which simplified procedures for implementation of the promises legalized by the 1917 Constitution. The next administration, under President Plutarco Elías Calles (1924–1928), continued the redistribution, and a high point was reached in the one-year administration of Emilio Portes Gil in 1929. The years from 1930 until 1933 saw a drastic decline of redistribution in the Pascual Ortiz Rubio administration, followed by an improvement with Abelardo Rodríguez's administration. After the enactment of the Agrarian Code in 1934, the Lázaro Cárdenas administration (1934–1940) distributed more land than in all the previous years of the reform. A peak was reached in 1937, when over five million hectares were redistributed, but a sharp decline followed in 1938 and 1939, probably the result of the increasing scarcity of haciendas for expropriation and administrative inability to cope with the large amount of land being redistributed. This resulted in disputes about the eligibility of land recipients, expropriation exemptions, and the accurate measuring of property lines.

The administration of Manuel Avila Camacho (1940–1945) continued land redistribution. The largest and most abrupt drop in redistribution occurred in 1945, when the administration concluded that large private holdings with arable lands were becoming extremely scarce (Whetten 1948:129). Redistribution continued on a reduced scale under Miguel Alemán (1946–1952), who distributed over five million hectares (Padgett 1966:193).

Alemán's administration also made a number of changes in the agrarian codes and reinstated the legal proceedings called *amparo* for granting exemptions from redistribution, as well as other changes intended to slow down and rationalize the land-granting procedure. The *amparo* was designed to prevent the expropriation of land and water from small property holders by landless agriculturalists. Large landholders were not affected. The changes in the land-granting procedures raised the minimum ejido holding to ten hectares of irrigated land and twenty hectares of seasonal land. The rationale was to prevent the further proliferation of small uneconomical plots, or *minifundios* (Padgett 1966:193).

The next administration, under Adolfo Ruiz Cortines (1952–1958), distributed an additional 3.5 million hectares, but land for redistribution was by then becoming extremely scarce. Ruiz Cortines emphasized the continued importance of small private landholding, or *pequeños propiedades,* along with ejido tenure as the best way to raise rural living standards. Part of the land that began to be distributed at this time came from the opening up of new lands through irrigation and reclamation projects. By the end of the Ruiz Cortines administration, there were 924 new land colonies occupying 6.2 million hectares (Padgett 1966:195).

The total amount of land redistributed in Mexico from 1916 to 1945 as a result of the agrarian reform was 30,619,321 hectares, which was allotted to 1,732,062 individuals. This hectarage represented 15.5 percent of the land area of the Mexican Republic (Whetten 1948:124). In the period between 1945 and 1958 another 13 million hectares were redistributed, and Adolfo López Mateos (1958–1964) added another 16 million hectares, which brought the grand total to 59.6 million hectares by the end of 1964 (Padgett 1966:195). By 1980, 2.3 million *ejidatarios* held about 80 million hectares of land (Susan Sanderson 1984:99).

Despite the long and extensive agrarian reform, which provided land for 50 percent of all those who now have access to land, the Mexican campesinos are faced with continuing structural problems. The vast majority of ejido holdings are less than five hectares, as are the more than 2.5 million *minifundio* holdings. Much of the land is of poor quality and is not irrigated. Price fluctuations, environmental hazards, exploitation by marketeers, and difficulty in obtaining credit or technical assistance have resulted in harsh conditions for most. Malnutrition is a serious problem in many rural regions, especially among children under ten.

Minifundio holdings are concentrated in the densely populated areas of the central and south central highlands in the states of Zacatecas, Jalisco, Guanajuato, México, Morelos, Hidalgo, Tlaxcala, and Puebla. The situation is exacerbated by the lack of irrigation water for a large percentage of the *minifundios.*

As a response to high population densities and a proliferation of *minifundios,* the government began to develop areas of high agricultural potential, namely, the large river basins and surrounding areas. Since 1946, these projects have had only limited impact on overall agrarian conditions, although spectacular developments have been achieved in some regions. The largest projects and their administrative river basin commissions are the Fuerte, the Lerma-Chapala-Santiago, the Tepalcatepec, the Balsas, the Pemco, the Papaloapan, and the Grijalva (Barkin and King 1970). These projects began in the middle and late 1940s and continue to have an effect on the rural population in the main project areas and surrounding regions. They were planned to relieve land pressure and open large new areas for resettlement and resource development, but accomplishments usually fell short of official projections.

The original ownership of both land and water rests with the Mexican state, which has the power to transfer ownership of land to private individuals. Most surface water, groundwater, mineral deposits, and the subsoil are under the direct ownership of the state and are inalienable and unalterable. Individuals and groups can, under specific conditions, however, be given the rights to exploitation. Water, according to Article 27, includes territorial seas, lagoons, estuaries, inland lakes connected to streams with permanent flow, rivers and tributaries from the source to the sea, and the streams that are constant or intermittent in flow when they form national and state boundaries or flow from one state to another or cross a national boundary. Other water is considered part of the property on which it is found or through which it flows. If this water is part of two or more properties, its use is considered part of the public welfare and is therefore under state control (Whetten 1948: 117–118). Since very few rivers do not connect to other rivers and bodies of water that eventually flow to the sea, virtually all surface water is under ultimate state ownership and administration.

Groundwater is part of the subsoil and is separate from both land and surface water. There is a section of the government water ministry that deals with the right to explore for and to use the groundwater resources (Secretaría de Recursos Hidráulicos, Dirección de Aguas Subterráneas). Private property, therefore, is defined as the right to "own" the surface of the land, but does not include the right to use water from surface resources or to dig for groundwater.

It is clear that the separation of the land surface and water into administratively distinct resources to which individuals and groups can have access is a variable that has influenced the structure of groups and the nature of action on the local level. The resources by themselves do not determine structure and use, but the institutionalization of access, as defined by the state and its institutions, does set the parameters for local-level social organization and subsequent action.

Article 27, in the interest of public welfare, imposed limitations and conditions on the transfer of property redistributed after the Revolution. The intention was to subdivide large holdings and to develop the farming of multiple small holdings leading to the creation of new agricultural communities. The law originally was intended to provide adequate land and water for the new agriculturalists. Communal property was legalized as a form of collective land tenure for centers of population and became the basis on which villages, communities, small towns, or population centers could petition for the restoration of previously held lands or for the grant of new land (Whetten 1948:120–130). Communal or ejido tenure was actually a restoration of the tenure system practiced prior to the Porfirian period and was therefore the accepted form of tenure in most rural areas. Individual holdings continued under very specific limits regarding maximum size. These limits were based on geographical divisions, type of agriculture, the availability of irrigation water, the location of the land in relation to a village petitioning for an ejido grant, and the availability of federal land. Private holdings within a seven-kilometer radius of the center of a petitioning village were subject to seizure unless their area did not exceed one-hundred hectares of irrigated land or two-hundred hectares of unirrigated land. The exclusion limits for expropriation were based on the following conversions: each hectare of irrigated land equated two hectares of unirrigated land, four hectares of pasture land, or eight hectares of woodland in barren parts of the country. Much larger holdings were permitted for cattle, cotton, bananas, coffee, and certain other products. The present-day ranches or former haciendas in the Tehuacán Valley are examples of individual private holdings that were permitted to continue, albeit on a much smaller scale than before.

The rural village was the basic unit to which the government granted new land or returned previously held land. A number of problems arose concerning the definition of the rural village or community, but in the Tehuacán Valley the grants were made to the *municipios* whose territorial divisions and populations were clearly defined and recognized. Problems in other regions often arose when individuals claimed to be an eligible "village" solely to receive a land grant. The government did not restore or grant land to individuals. Individuals could, however, share in a land grant by participating in the collective working of ejido land or by individual working of a small portion.

Expropriation of land and redistribution to the village was done by the federal Agrarian Department (Departamento Agrario), later changed to the Department of Agrarian Concerns and Colonization (Departamento de Asuntos Agrarios y Colonización—DAAC), which in 1977 was elevated to full ministry status as the Ministry of Agrarian Reform (Secretaría de la Reforma Agraria—SRA). The final decision to expropriate or to make a land grant always rests with the president of the republic, an indication of the highly centralized structure of the Mexican government. The initial petition for land is made by the village and submitted to the state governor, who refers the matter to a three-member Mixed Agrarian Commission. This commission, a branch of the Agrarian Department, determines the number of eligible recipients in a village and the amount of land that can be taken. The Mixed Agrarian Commission then reports to the governor, who has the authority to make a temporary grant, pending final approval from the president (Simpson 1937; Tannenbaum 1929; Whetten 1948:136).

The size of the grants was initially four hectares of irrigated or wet land or eight hectares of unirrigated land for each eligible participant, maximum. The local situation dictated how much land could actually be allotted. Revisions in the Agrarian Code made in 1943 increased the allotment to six and twelve hectares. No doubt, the revisions created inequities between earlier and later recipients of land.

Individuals, as participants in an ejido grant, were given a form of limited ownership that entitled them to farm the land during their lifetime. The use rights could be inherited, but the land could not be sold, mortgaged, or leased. Farmers could lose their right to a share in the ejido land grant if they tried to sell or otherwise transfer the property, married someone who also held land, were convicted of a criminal offense and sentenced to jail, became alcoholics or went insane, failed to take possession of the land, failed to pay taxes or meet the obligations imposed by a general assembly of landholders, or committed acts against the community that caused disorganization, confusion, and disharmony. These rules were in effect until 1943, when, because of abuses by local officials, landholders, and politicians that deprived many of their land, the law was changed. The revised law stated that failure to work the land for two consecutive years was the only grounds for losing the right to participate in an ejido grant.

The most significant aspect of the ejido land grant, in terms of the individuals who hold and farm the land, is the local-level organization of ejido landholders prescribed by the Agrarian Code. This ejido organization, together with the *municipio* government and local institutions such as irrigation organizations, are the sociopolitical structures within which all *ejidatarios* must operate. In terms of economic activities and the ability to cultivate sufficient corn for domestic needs and commercial profit, the ejido and irrigation associations are the most important organizations. The two operate independently but their memberships overlap, their leaders are often the same, and their activities are interrelated. Every ejido in Mexico is organized into a general assembly consisting of the entire membership of all participants in a land grant. The assembly elects two commissions whose purpose is to manage the financial, legal, and administrative functions of the ejido. The first and most authoritative is the ejido commission, or *comisariado ejidal,* consisting of three elected members and three alternates. The second is the oversight committee, or *comité de vigilancia,* also consisting of three members and alternates.

The general assembly is supposed to meet once a month, but once an ejido is in operation, the land subdivided, and titles issued, the meetings become much less frequent. Notice of a meeting must be given one week in advance, and a quorum consists of one-half the total number of assembly members or *ejidatarios* plus one. The members of the ejido commissions and oversight committees are elected for three-year terms by the general assembly. Members of both may be reelected by a two-thirds majority. According to the Agrarian Code, the general assembly, in addition to electing the ejido commission and oversight committee, also serves to modify, authorize, and rectify commission and committee decisions; to discuss reports; to request the intervention of agrarian authorities on matters involving the suspension of ejido members or depriving them of landholding rights; to issue rulings on how the ejido land should be used; and to submit all rulings to the Departamento de Asuntos Agrarios y Colonización for approval.

It is important to note that in all local rules, codes, and regulations there are clauses that dictate the legitimate involvement of the government in virtually all phases of local activities and organization. Government intervention, however, has been infrequent, and most acts and decisions submitted for approval are "rubber-stamped" as a matter of routine. But if needed, the structural prerequisites and pathways are built in for an immediate government takeover.

The three ejido commission members consist of a president, a secretary, and a treasurer, who must be residents of the *municipio* and literate. These members perform the active administration of the ejido, such as representing it in all external relations, overseeing the use of the land, implementing and controlling the division of collective land into individual plots, calling meetings of the general assembly, overseeing all work performed, and complying with and enforcing the rulings of the general assembly.

The oversight committee functions as an overseer of the ejido commission to ensure that all activities are in accordance with the Agrarian Code and that all other applicable laws, rules, and regulations are followed. The committee has the right and duty to inspect all books and can make reports to the general assembly. It can call special sessions when irregularities are encountered. Irregularities and infractions must also be reported to the Departamento de Asuntos Agrarios by the oversight committee if the ejido commission fails to do so.

Along with land redistribution, the agrarian reform also recognized the importance of granting water rights to the *municipios*. The ejido laws contained provisions for the distribution of irrigation water, but by the end of 1933 only a fraction of the *municipios* receiving land grants had also been given definite water rights (Simpson 1937:187). The relatively scarce water resources in large areas of Mexico have accounted for the lack of water rights to go with the redistributed land. The Federal Irrigation Law (Ley sobre irrigación con Aguas Federales) enacted in 1926 established the National Irrigation Commission (Comisión Nacional de Irrigación), which in 1946 was elevated to ministry status and renamed the Ministry of Water Resources (Secretaría de Recursos Hidráulicos—SRH). The SRH, an organization of engineers, lawyers, and administrators, worked in conjunction with the Ministry of Agriculture and Livestock (Secretaría de Agricultura y Ganadería—SAG), and the Departamento de Asuntos Agrarios y Colonización to develop and distribute irrigation water from all existing and potential resources. Consequently, the SRH was in charge of the administration and direction of all use, maintenance, and construction of physical facilities relating to both surface and underground sources of water, but had little overall impact on the availability of irrigation water in the Tehuacán Valley. In 1976, under José López Portillo, the SRH was combined with SAG and was renamed the Secretaría de Agricultura y Recursos Hidráulicos (SARH).

The federal judiciary, a highly centralized system, functions to regulate and litigate questions of individual access to public and private land and the use of natural resources. The judiciary is divided into a Supreme Court, federal circuit courts, and federal district courts. The important distinction to be made in regard to litigation over land and water is that only the Supreme Court can hear *amparo* proceedings. The lower courts have jurisdiction in criminal cases arising from a violation of federal laws or treaties, crimes committed by federal employees during their employment, and all crimes that prevent a federal department from exercising its power. Federal district courts hear civil cases arising over controversies concerning the application of federal laws, as well as cases to which the federal government is a party (Martínez 1968:220). The First and Second District courts in Puebla have jurisdiction over the part of the Tehuacán Valley where this study was carried out. The Tribunal de Primera Instancia in

Tehuacán hears most civil and criminal matters that do not involve *amparo* proceedings.

The Public Ministry of the Federation (Ministerio Público de la Federación) is a part of the executive branch, the Mexican equivalent of the Justice Department. It prosecutes all civil and criminal cases in the local courts involving the government and its institutions. The Ministerio Público, with a regional office in Tehuacán and a representative in each *municipio*, can also enter local disputes to try to resolve differences between individuals, institutions, and organizations before they are taken to the courts.

Another important part of the postrevolutionary institutional structure is the local *municipio* and its elective offices. The *municipio* receives its legitimacy from Article 115 of the 1917 Constitution, which defines the *municipio libre* as the basic local administrative unit. The Constitution specifically excludes the district-level political boss or *jefe* as an intermediary between state and locality.[4] Article 115 goes on to say that the *municipios* shall freely administer their finances and shall be financed by taxes imposed by the state legislature, which shall be sufficient to cover the *municipio*'s necessities (Secretaría de la Presidencia 1970:47). The *municipio* is therefore at the end of a highly centralized political structure connecting with the executive branch of the federal government through the state legislature and the governor's office.

The most important *municipio* officers are the *presidente municipal*, treasurer, secretary, and eight *regidores*, or council members, who are elected by direct popular vote for a three-year term. An individual cannot hold successive terms in the same office (Secretaría de la Presidencia 1970:48). The salaries and operating budgets of the *municipio* and its officeholders are extremely low. In 1973, a *presidente municipal* made less than ten pesos a day.[5] The amount of money at a *presidente*'s disposal, and therefore the range of things one can hope to accomplish, is very small. For anyone eager for accomplishments, the work is strenuous, sensitive, and frustrating. The power, effectiveness, and impact of the *presidente municipal* are very much dependent on who holds the office. For some, the office has been a passive service of civic obligation; for others it has been a period of action and change that has led to other positions within the government.

The *presidente municipal* instigates innovation and improvement, arbitrates personal conflicts, acts as a broker or advocate when seeking funds or materials from outside sources, and may be instrumental in seeking cooperation within the *municipio*. In the small *municipios* of the Tehuacán Valley, the *presidentes municipales* have been concerned with the installation of potable water, street surfacing, sewers, and the construction of schools and health facilities. In the case of major construction, it has been the *presidente*'s success with state and federal officials that has made the difference in getting both financial and technical help.

A third institutional element in the local *municipio* is the "grass-roots" level of the highly centralized and ever-present Partido Revolucionario Institucional (PRI), the political party in power since the Revolution. The PRI has committees in the valley *municipios* and is organized around the leadership of the local deputy (*diputado local*) to the state Chamber of Deputies in Puebla. Although it is difficult to say how much power *diputados* can exercise, it is clear that, together with the party apparatus, they have considerable influence in the selection of the candidates for local *municipio* offices. Other political parties do exist, but they are a minority and have until recently rarely won in *municipio* elections. The *diputado local* is extremely visible and frequently heads village parades, gives speeches, and introduces visiting politicians. The *diputado* does have a considerable amount of influence within the party structure and may also be able to influence decisions made by the governor and federal officials in the allocation of funds, goods, and services to the local *municipios*.

POSTREVOLUTIONARY CHANGES IN THE TEHUACÁN VALLEY

After the Revolution, agrarian reform was implemented very slowly. The major haciendas in the valley were broken up over a twenty-year period and the land was redistributed as ejido grants at different times. There are no published accounts of precisely when particular haciendas were seized and how much land was given to particular *municipios*. Haciendas near the location of this study include La Huerta, San Andrés, Buenavista, Santa Cruz, San Francisco, and Venta Negra; these were redistributed between 1928 and 1939. La Huerta and Buenavista still exist on a reduced scale, but the other four do not.

When the haciendas were broken up, only the land was redistributed; in many cases, the water rights were retained by the former hacienda owner. As a result, the *hacendado*, with greatly reduced landholdings or none at all, continued to control the most valuable component for agricultural production. According to Whiteford and Henao (1979), the *hacendados* were able to keep their water rights by bribing government officials to classify their lands as nonirrigated.

They then sold water at a great profit to the newly formed ejidos. The *ejidatarios,* however, were petitioning for the water rights and also threatened to take them by force. The 1930s and 1940s became a period of intense conflict over water among individuals, associations, and communities.

General Lázaro Cárdenas was the driving force making the campesinos' dreams of regaining land a reality. Yet the confiscation of the haciendas and the formation of ejidos were not always coordinated with the redistribution of water rights. In some cases, the ejidos were unable to gain rights to water to irrigate their fields, and without the water the fields could not be cultivated.

Altepexi

The postrevolutionary history of land reform and redistribution in the *municipio* of Altepexi began in 1914 and was followed by a lengthy process of arbitration.[6] On December 29, 1933, President Abelardo Rodríguez signed a provisional ejido grant, but the agriculturalists had to wait for the official act of possession and surveying (Acta de Posesión y Deslinde), signed on November 20, 1938, to work the land. This was the time when most of the ejidos in the Tehuacán Valley received official sanction to begin operation. The final presidential decree legalizing the ejido grant was not issued until 1953, and the local ejido officials did not receive complete and legal documentation until 1974. What follows is a brief history of the ejido grant made to Altepexi, with a description of the land quality and its intended use.

On March 28, 1915, Altepexi first petitioned the governor of the state of Puebla for the restitution of land and water under the control of the haciendas of Santa Cruz, Buenavista, San Francisco, and Venta Negra. The basis for the petition was the 1790 communal tenure and the traditional right to two *caballerías* of land.[7] One *caballería* had been legally purchased and the other was donated by the king of Spain. Furthermore, the community had rights to forty-eight hours of water per month from the natural spring San Andrés, and to one nightly furrow (*surco*) of water, which was adequate for household needs. The community also claimed title to another *caballería* of land, but this title had neither been confirmed nor recorded. The petition further claimed that Lieutenant General Prisciliano Ruiz, when he marched through the region in late 1914, had provisionally donated an unspecified amount of land, part of the flow from La Taza spring, and the total flow from Cozahuatl and Atzompa springs.

The Local Agrarian Commission (Comisión Agraria Local) began working on the petition by the end of May 1915 and sent official letters to the owners of the Venta Negra and San Francisco haciendas to inform them of the claim made by Altepexi. The commission did not make any recommendations, because the titles to the lands claimed by Altepexi could not be legally confirmed.

On November 30, 1917, Altepexi again petitioned the governor and asked the Agrarian Commission to disregard the previous petition, because the earlier tenure could not be confirmed. The new petition requested a new land grant under the Decree of 1915 and Article 27 of the 1917 Constitution. Instead of restitution, Altepexi asked for an outright grant of sufficient fertile land, pasture, and water to satisfy the agricultural needs of the population.

No action was taken on the second petition until May 1925, when the governor of Puebla, on the recommendation of the Local Agrarian Commission, made a provisional grant to the *municipio* of Altepexi. The grant approved the terms of the military expropriation made by General Ruiz and stated that as of June 30, 1925, Venta Negra must surrender 977 hectares of arable land. Furthermore, Altepexi was granted an unspecified portion of the flow from La Taza spring and the entire flow from Atzompa spring, previously controlled by the hacienda. At the time, this amount of water was said to be enough for domestic needs and for irrigating about 250 hectares of corn.

Before the grant was made official, the National Agrarian Commission, with power over the local delegations, ordered that a new census be made of the petitioning *municipios* to make a better estimate of agricultural needs. The grant given to Altepexi, for instance, was based on data from the 1918 census, in which the population was set at 1,172, of which 345 were judged eligible to participate in an ejido grant. Many *municipios* were asking for a new census to show population increases and more eligible individuals.

The new census, made during 1926–1927, contained current and more complete data. It showed that Altepexi had 1,585 inhabitants, 777 of whom were eligible to participate in an ejido grant. The census also showed that the *municipio* consisted of a total of 896 hectares, of which 87 were occupied by nucleated residence units, 328 were cultivatable with irrgation, 477 were hilly pastures not suitable for agriculture, and 4 were occupied by a textile mill. The arable land was subdivided into small plots no larger than three-quarters of a hectare and cultivated individually by the inhabitants. The pasture was used communally. For irrigation and domestic purposes the *municipio* had been using a portion of the flow from La Taza according to the terms of the grant made by the governor. No land had been effectively expropriated from the haciendas.

The census went on to show that the inhabitants of Altepexi were primarily agriculturalists and that most of them worked as laborers on the nearby haciendas. The climate was warm and rainfall inadequate for agriculture without irrigation. The main crops in Altepexi and the neighboring *municipios* were corn, wheat, hops, beans, tomatoes, and sugarcane.[8]

The five haciendas in the area had a total extension of 7,710 hectares, of which only 890 hectares were classified as irrigated. The source of the irrigation water was natural springs and *galería* systems. No data have been found to indicate the amount of water that was available from these sources.

At the time of the census, the Local Agrarian Commission had available almost eight-thousand hectares for potential ejido grants to satisfy at least twelve petitions for "adequate" ejido grants of both land and water. All of the major *municipios* in the valley had petitions for land considered by the commission for a final recommendation to the governor and eventually to the president. Approximately one-hundred hectares were permitted by law to remain under the ownership of the *hacendados,* who were reclassified as *pequeños propietarios.* This was the maximum size that the agrarian reform laws allowed for private tenure. Water from the natural springs was subject to expropriation, but the *galería* systems remained the private property of the original owners.

From 1926 to 1929, the Local Agrarian Commission examined and made recommendations concerning the eligibility of the ejido petitions, determined the number of eligible individuals per petition, and estimated the amount of available land and water. The land reform process also took into consideration the petitions for noneligibility by the hacienda owners, their heirs, and relatives.

The provisional ejido grant made to Altepexi in 1925 stated that the *municipio* had demonstrated its eligibility to receive land according to the agrarian reform laws. Because there was inadequate evidence to demonstrate conclusive prior ownership of land and water, which had been illegally taken after June 25, 1856, Altepexi was, according to the agrarian reform codes, eligible for sufficient land to satisfy the needs of its population. The Agrarian Commission, using the data from the most recent agricultural census, found that 656 persons were eligible to participate in the land grant.

The agrarian reform limited the amount of land that each person could hold to no more than 3 hectares of irrigation land, 6 hectares of rain-dependent land, or 8 hectares of pasture. Using these criteria, Altepexi received a total of 3,707 hectares from four of the five haciendas described above, reducing the haciendas to the maximum size of 100 hectares each. The new lands were classified as 388 hectares for irrigation, 1,250 hectares for rain-dependent farming, and 2,069 hectares for pasture.[9]

Using the guidelines of the agrarian reform, we can calculate the number of individuals that the grant accommodated. The 388 hectares of irrigation land were adequate for 129.33 individuals, the 1,250 hectares of rain-dependent land for 208.33 individuals, and the 2,069 hectares of pasture for 258.63 individuals. The total number that could have been supported by the ejido grant was 596.29, leaving 59.71 eligible people without land. The 477 hectares of communally held pasture was entered into the land reform calculations and judged sufficient to satisfy the remaining petitioners. The grant did not specify how the lands should be distributed among the eligible individuals, but left it up to the ejido commissions in each *municipio*.

Water was granted to the ejidos separately from the land. The fact that Altepexi had received 388 hectares of irrigation land did not mean that sufficient water was included. The land was only classified as irrigated in order to calculate the number of participants in the grant. For all practical purposes, both the irrigation and rain-dependent land, a total of 1,638 hectares, needed to be irrigated to make cultivation possible, and the pasture was considered unsuitable for agriculture.

La Taza spring, located just north of Altepexi, was the source of water for the ejidos in the *municipios* of San Gabriel Chilac, San José Miahuatlán, San Francisco Altepexi, and San Juan Ajalpan. The 250 lps flow from La Taza had to provide irrigation water for six ejidos and satisfy the domestic water needs of the four *municipios*. The amount of water was inadequate, but no other source was available for distribution.

The presidential decree dividing the water from La Taza was issued on November 15, 1939, and finalized on June 14, 1947. This meant that the ejidos and the *municipios* could begin using the water at the same time that the ejido land grant was formalized by the Acta de Posesión y Deslinde. The flow from La Taza was divided evenly, giving half the flow to Miahuatlán and Chilac and the other half to Altepexi and Ajalpan. The water for Altepexi and Ajalpan was to be divided between three ejidos, two in Ajalpan (San José Buenavista and Teopuxco) and the one in Altepexi. A division in thirds gave each 41.6 liters per second. Not counting the amount used for domestic purposes, the flow only sufficed to irrigate, at most, sixty hectares of corn. This represented 3.6 percent of the total agricultural land grant to Altepexi, not including the pasture.

Almost twenty years after the end of the Revolution, the net result of the land reform for the inhabitants of Altepexi was a large land grant and practically no water. The *galería* systems that could provide valuable additional water were still under the control of the original owners, the dispossessed *hacendados* and their heirs. The alternative for the ejido landholders was to purchase existing *galería* systems or to construct new ones; the *ejidatarios* in Altepexi did both.

Chilac

In San Gabriel Chilac the union of cane cutters remained a dynamic organization (with the dynamic leadership of Melitón Ramírez), despite its problems on the valley haciendas. Through the union, the landless residents of Chilac started the lengthy process of petitioning for land held by neighboring haciendas.

After the Revolution the haciendas near Chilac were gradually forced to relinquish part of their land. In 1938 Hacienda San Andrés had most of its 5,177 hectares divided between six communities; Chilac received 1,010 hectares, but only after lengthy requests to the president and the governor. Although a large amount of the land received had formerly been irrigated by the hacienda, the ejido formed in Chilac received no water from the hacienda, much to the frustration of the *ejidatarios,* who realized the land had little value without water.

San Marcos Necoxtla, the community located north of Chilac, had at an earlier date lost to Hacienda San Andrés its right to water from springs rising inside and just outside the village. San Marcos was also denied its request for water from the hacienda.[10] After a series of unsuccessful petitions to the governor for water, the ejido of Chilac offered to sharecrop the land it had received. The hacienda owners would provide the water, seed, and oxen, the *ejidatarios* would provide the land and the labor (Carta de la Agrupación Agraria de San Gabriel Chilac 21 de julio 1932). The request was turned down by the owners, who had maneuvered to have ejido lands declared nonirrigated. Realizing they would eventually lose the water either to the ejido in Chilac or to San Marcos, the owners sold part of their water rights to a water association in 1939. The water association, Sociedad Chilac-Tlacoxcolco-Tititlán, was made up of men from Chilac who had not participated in the ejido activities as well as a few men from Altepexi. In 1948 the owners of the hacienda decided to sell most of their remaining water and put 192 hours of water up for sale. The ejido desperately tried to raise the money but could not get any government assistance. The same water society that had bought the first water from the hacienda bought the second group of shares.

The ejido of Chilac also received 393 hectares from Hacienda Venta Negra in 1931, when 1,697.5 of the hacienda's 2,015 hectares were divided between communities in the region. In 1933 the ejido petitioned for water from the spring Atzompa, previously used to irrigate the lands it had received from the hacienda. The governor of Puebla had provisionally given the water to San José Miahuatlán, which had been buying water from Hacienda Venta Negra. However, since no ejido existed in San José, the governor retracted and gave the water and land to San Francisco Altepexi. In 1936 the ejido in Chilac finally received a presidential decision granting them the water from Atzompa. In December of the same year an armed band of men from San José took over the irrigation canal and the dividers that sent water to the respective villages. The Chileños complained to authorities in Tehuacán, who sent troops, capturing six men while the rest escaped. In the ensuing court case, San José made new requests for water. In 1945 a water association was formed in San José that bought rights to twelve days of water from the Hacienda Venta Negra. The ejido of Chilac remained with eleven days and sixteen hours monthly of water from San Agustín of the spring Atzompa. The quantity of water, 36.7 lps, was only sufficient to cultivate 48 hectares of land and was divided between 144 *ejidatarios*. Unfortunately, in about 1954 Atzompa went dry and the Chilac ejido lost its last water.

In Chilac, the conflict over water between families that already controlled land and water and those that did not suggests class-based antagonism and competition. The cane cutters' union did in fact have a class base and had openly mobilized landless agricultural workers to challenge the status quo at home as well as on the plantations. When the dynamic leader of the Chilac union was shot in Tehuacán, union members blamed his death on two groups. The first was the neighboring community of San José Miahuatlán, which had been in conflict with the union over water. The second was the network of valley caciques, who controlled the villages in the valley and who were threatened by the agrarian movement and the formation of the ejidos. The shooting of their leader frightened the *ejidatarios* and quelled the agrarian movement.

The struggle for control of water and land in the Tehuacán Valley had many contestants. For most of the indigenous population, the major threat to their control of land and water came from the Spanish and later the urban-based mestizos. In many ways they have managed to restrict the control of most of the vital resources to those who live in the valley. Nevertheless, the pressure from the outside for land

and water is constant and growing, especially with the expansion of agribusiness.

Competition for land and water between communities as well as between sectors or even classes within communities has emerged over the years as a dominant fact of social life in the valley. Conflict generated by the competition for water and land has intensified. The consequences of this competition and conflict for everyday life are immense. The contemporary nature of these processes is examined in the next chapter.

5.

COOPERATION AND DIFFERENTIATION

THE AFTERSHOCKS of the Revolution resounded throughout Mexico, lasting well into the 1940s. In the Tehuacán Valley, the late 1930s and early 1940s was a period of transformation marked by the formation of new ejidos, battles over water rights, development of new sources of irrigation water, and the gradual emergence of political stability. Tehuacán-based mercantile companies, battered during the revolutionary period, lost their power and were never again a force in the valley. Rapid urbanization sparked the expansion of new markets for food crops, and indigenous entrepreneurs from the Tehuacán Valley created elaborate trade networks for transporting corn, *elote,* garlic, and tomatoes to the growing metropolitan markets of Mexico City, Veracruz, and Puebla. Capitalist agriculture replaced subsistence production in many parts of the valley. The population began to expand rapidly, putting new pressures on resources. Competition for land and particularly water resources intensified. Prices for both rose rapidly.

These pressures contributed to the social dynamics that led to the creation of cooperative associations, called *sociedades explotadoras de aguas,* which financed the construction of the *galería* wells and ultimately expanded the available water supply. Entirely private enterprises, the associations received neither credit nor technical aid from the national government.

The shift to commercial agriculture, the expansion of water resources, and the pattern of resource ownership were all factors in the transformation of the social structure and interpersonal relations throughout the valley. In this chapter we describe the emergence of irrigation associations as social institutions within an expanding capitalist economic context and explore factors that contributed to their creation and perpetuation. We then explore whether the associations generated equity within the communities, or became vehicles for a small sector's accumulation of wealth and power.

COMMUNITY STRUCTURE AND THE EMERGENCE OF CORPORATE COOPERATION

Before we begin, we would like to put the chapter in a broader comparative context. Social relations in "peasant communities" have been characterized by some scholars as atomistic. "An atomistic-type society . . . is a society in which the nuclear family represents the major structural unit and indeed, almost the only formalized social entity. Interpersonal relationships outside of the nuclear family are characterized by contention, suspiciousness, and invidiousness" (Rubel and Kupferer 1968:189–190). These characteristics, which are passed from one generation to another, tend to inhibit cooperative activities. "When attempts at large-scale organization are made, they often fail because people are unready or unwilling to collaborate and cooperate" (Honigmann 1968:221). George Foster (1967:136) has suggested that the lack of cooperation is a rational response, given peasants' world-view that more good than already exists *cannot* be produced by cooperation, and that cooperation can only threaten the balance and create risk. The atomistic attitude has been cited as a factor in preserving backwardness and impeding development (Banfield 1958; Rogers 1969). A more materialistic explanation traces the lack of cooperation to the forces of capitalism, which generate proletarianization leading to individualization and contrived competition (Alexander and Alexander 1978; Barnett 1984; Biswas 1986; Coward 1979; Freivalds 1972). During the early stages the process makes local-level political and economic mobilization difficult.

In contrast to the peasants of the atomistic communities described by the authors cited above, many of the residents of the Tehuacán Valley participate in a variety of cooperative organizations.[1] There is not only an economic rationale for participation but also community pressure to cooperate in some of the larger canal associations. The result is the elaborate decentralized system of different types of organizations that handle distinct aspects of water management in the valley: canal associations, spring associations, community water organizations, associations that control floodwater in the arroyos, and *galería* associations.

Spring associations (*sociedades de manantiales*) are similar to *galería* associations except that members own shares of springwater instead of well water. These associations are particularly important in San José Miahuatlán, where the three major springs—Atzompa, Cozahuatl, and Coyoatl—are owned by associations. The ownership of water, especially Coyoatl, was the key to power in San José. When Coyoatl ran dry, *galerías* were built. In Altepexi, an association owns a part of the water that comes from the spring La Taza.

Another type of association is the *sociedad de canales,* canal association. Canal associations are made up of individuals who own land along the canals that carry water to the fields from the water dividers. In Chilac, for example, there are thirteen canal associations. One association, the Sociedad del Canal Reforma, has more than seven hundred members. The major canals have a series of "branches," or *ramas,* which take the water out to the fields. Each of these branches also has its own organization. The associations have their own charters, officers—president, secretary, treasurer, and messengers (to tell members about meetings or work projects)—and captains for each of the subcanals. Members are required to work either on the cleaning or the maintenance of the canal or to pay a fine.

The floodwater associations (*sociedades de aguas broncas* or *sociedades de zanja*) have built channels to tap the floodwaters that periodically rush down the arroyos. Although their organizational structure is similar to that of the other associations, they must be able to mobilize their members rapidly to take advantage of the short periods of the arroyo flow.

Associations developed to control floodwater are composed of people who own land that can be irrigated by water from the canals that are linked to the floodgates that are placed in the dikes to protect the fields from the arroyo. The water from the arroyo is important for the fertile soil it carries as well as for the moisture it provides. These forms of organization probably predate the arrival of the Spaniards.

In the final type of association—the *galería* association—the cooperation necessary to finance and to develop the *galería* wells stems from traditional organizational principles used by other types of water-management organizations in the valley. Yet the *galería* associations are different in that criteria for membership are not based on living or owning land in a specific area. Instead, *galería* associations are formed by friends and kin who feel they can work together, thus reducing the potential for conflict and adding an element of dependability. The pattern tends to cut across economic strata within communities, but because of the necessity for long-term investment, the most wealthy members of the communities have a greater opportunity to participate. We elaborate on the consequences of this process later in the chapter. Most significantly is that the process tends to reduce an individual's perception of class differences during initial stages of the association. As the costs mount and the returns are delayed, the poorest members often cannot pay the quotas and have to sell their shares. In some cases, this does generate resentment and antagonism.

Although people join *galería* associations hoping to make a profit, there is considerable risk. Furthermore, many are reluctant to incur a long-term financial obligation. People often pressure friends or relatives to join and invest to make a venture possible. If one's friends cannot develop a resource, it is assumed that other groups will take advantage of the situation and increase their wealth. In this sense, there is a perception of limited good, but to a degree it is seen from a group perspective, which in turn generates pressure to cooperate.

People may belong to a number of irrigation organizations at the same time. In Chilac some people belong to three or four *galería* associations, several flood-water associations, and several canal organizations. In all of these cases, they work together because they all benefit when the irrigation system is well coordinated. There is social pressure to resolve conflicts within the group and not to take them outside the community. This elaborate system of overlapping decentralized organizations provides the bases of indigenous control of water in the valley.

The irrigation organizations are a response to the social struggle for resources. Through them the members have coordinated action, creating access to resources as well as to power. The fact that the indigenous population in the Tehuacán Valley formed a multitude of corporate organizations can be traced to several factors that created a favorable environment for the emergence of cooperative efforts. The major communities have relatively large populations, ranging from four thousand to seven thousand people. Atomism has been attributed to small populations. More important, there may be a relationship between atomism, the rate of population growth, and the rate of expansion or contraction of the resource base (Spielberg 1968:211). In the Tehuacán Valley the *galerías* have significantly expanded the resource base, generating greater agricultural production and commerce. Certainly, in a differentiated large population incorporated in a capitalist system, competition and alliances can be expected to generate the formation of corporate groups. Capitalism gives survival advantages to those who, for whatever reasons (control of capital, labor, or markets), can coordinate activities. This in turn enhances the ability of group members to ultimately exploit the unorganized.

From the individual's perspective, cooperation makes participation possible, because none of the agriculturalists can finance *galería* construction alone. As a group, they share the risk and can pool enough money to complete the project. Returns on investment are potentially high, but so is the possibility of failure, which can be as high as one out of every two wells dug.

Ideologically, there are no sanctions against speculation and coordinated economic activity. This view may be traced to the tradition of irrigation that has been passed on through the generations. Irrigation in the valley has always required cooperative efforts, without which agriculture would have been extremely difficult. As a result, people in communities where irrigated agriculture has been practiced for centuries may be ideologically more open to opportunities where cooperation can generate new resources. Many of the traditional management principles of the community water system used in the Tehuacán Valley form the bases of the organizational structure of the irrigation associations.

For centuries in the valley, the distribution of water and the management of the irrigation systems were handled by a single community organization. The *galería* associations that emerged during the period of capitalist expansion retained many of the traditional responsibilities and principles of community organization, but included an important dimension of capitalist corporate structure: an opportunity for unequal investment and ownership. From the traditional principles of water organization evolved the democratic structure of the associations, the annual change of leadership, and the rotation of leadership between members. Water is distributed by turns throughout the month. The system of shares, measures, canal use, and other terminology was transferred from the community system to the *galería* system. Equally important, the maintenance and cleaning of the systems is shared by members and is obligatory; fines are assessed if obligations are not fulfilled.

Another characteristic taken from the structure of community organization is the custom of having a patron saint and celebrating religious holidays. Each society has a patron or a saint. Even the societies that are not under the advocation of a religious figure celebrate the day of Saint Croix on May 3, a festivity associated with work and water. This celebration includes a mass and a meeting for members. Mass is celebrated, then followed by fireworks, mariachi music, and dancing. Often the festivities take place by the *galería*, although afterward members go to the home of the elected head of the festivities committee for a meal.

DIFFERENTIATION AND CONSOLIDATION OF WATER RESOURCES

Although the *galería* associations made it possible for the local inhabitants to develop water resources without outside capital or experts, they involved only a segment of the population. In the valley there are more than 20,000 agriculturalists, of whom 10,219 are landless laborers and 7,425 own some type of irrigation water (SARH 1982). In the only community for which we have detailed information, San Gabriel Chilac, 22 percent of the households that are primarily dependent on agriculture for a living own at least six hours of irrigation water, 33 percent of the households that own agricultural land have six hours of irrigation water, and only 3 percent of the households that depend on agriculture for their livelihood but do not own land have water. At the same time, 12 percent of the households control 53 percent of the water. We will discuss these figures in greater detail later, but they do indicate that the water is not held equitably by the member agriculturalists.[2]

Galería associations are usually formed by a small number of good friends, often relatives or neighbors, who decide on the location for a well. The location may be selected with the help of a *visionario,* a local specialist in finding water, or by figuring out where water might be by studying the locations of other wells and springs. The group decides jointly whom to invite into the association. Generally, memberships are based on friendship ties, but dependability and fiscal viability are also considered. Failure to include a person who feels entitled to be invited for reasons of friendship or consanguinity can generate envy and hostility. One of the common explanations for a well's running dry is that the failure was caused by the envy of the people who were unjustly excluded from membership.

Although membership usually is based on informal friendship networks, other criteria are sometimes used. Membership in one association in the valley is based on membership in a political party, for example. The associations tend to fragment large communities, but they only reflect the informal patterns that are already established. Membership often includes people with a range of incomes, although the poorest peasants, even if invited, seldom have the capital to participate. There is one association, however, known as the association of the poor, in which a conscious effort was made by the founders to exclude more wealthy peasants for reasons we will discuss later. And in a few cases, ejidos have attempted to construct wells, but internal dissension has often limited their success.

Despite the corporate nature of many of the valley communities, membership in the *galería* associations is not restricted to residents of the same community. Although membership usually follows community lines, many exceptions provide important linkages between communities. These linkages evolve in many ways. First, an association that needs to dig holes on the land of a nonmember, from the same or a different community, can offer a share instead of cash remuneration. Second, the need for technical help generates intervillage linkages, since welldiggers and bookkeeping specialists do not reside in all communities. Working for *galería* associations based in other communities, these individuals may accept shares for payment. Third, the major communities of the valley are highly stratified. In some cases, associations are formed by wealthy campesinos from the different communities. For example, associations in San José Miahuatlán have members from Chilac and San Sebastián; and associations in Chilac have members from San Pedro, San Pablo, and Altepexi.

Migration has also influenced the membership in irrigation associations. Large-scale migration from the valley to Mexico City, Veracruz, the city of Tehuacán, and other urban areas has occurred since the mid-1960s. In many cases, the migrants have maintained close ties with their home communities, and some have used the income earned in the city to buy shares in irrigation associations. For example, an association was formed in Tehuacán by migrants from three major valley communities. The association also included a few people who still lived in those communities. The Tehuacán residents were interested in the water as an investment rather than for use on their own land. This pattern reflects a new dimension in the valley's integration in the capitalist sector: investment in agriculture by nonagriculturalists, such as truckers who live in the valley. These investors are contributing to removing the means of production from the control of the peasants.

The original cost of a share in an irrigation association varies considerably, depending on the number of members, the speed of construction desired, and the starting time. Payments used to finance well construction, based on the number of shares or hours of water a person owns, are usually collected every month, but may be collected each week, depending on the group decision. In 1977, a share cost between 80 and 120 pesos. The purchase of shares is viewed as an investment. Before water is found, a member can usually sell a share if the possibility of finding water is high. Once water is found, the value of shares increases dramatically. In 1977, the value of a share in an established water association varied considerably, depending on the amount of water produced by the well. For example, a share in the association Agua Escondida of Chilac cost 100,000 pesos, and in the association of San Isidro of Ajalpan, 180,000 pesos. Agua Escondida and San Isidro are the most productive wells in the valley, and the cost of their shares reflects that productivity. Shares in other societies that produce much less water may cost as little as 1,000 pesos.

The *galería* associations have been used by wealthy members to accumulate more wealth, often at the expense of less affluent members. The increasing costs of well construction or personal financial problems cause some people to sell their shares. Association bylaws stipulate that shares must be offered to other members before they are offered to nonmembers. One consequence of this rule is an increased concentration of shares and, subsequently, of water resources in the hands of the wealthier peasants. This is particularly true for wells that promise large amounts of water but that take a long time to construct. Investments of this nature are facilitated if individuals already have an assured income from water sales in other associations or other sources of income (Whiteford and Henao 1979).

A CASE STUDY OF ACCUMULATION

In the following case, we describe how the process has led to the concentration of control within the associations, and the importance of water resources in determining a person's position within the community. Our analysis focuses on the community of Altepexi, a bustling agricultural town of 12,500 (1980 census). Altepexi has a long history of conflict with nearby haciendas over water resources. Ejidos were formed in 1932 with land from the major haciendas. The major irrigation associations were formed between 1946 and 1973.

Because the volume of water flowing from each *galería* system is finite, each *socio* (member) receives an amount in proportion to his or her financial or labor investment in the *galería* purchase or construction. The flow of each system is divided into shares called *acciones,* which are fixed units of time. Each *acción* or *agua* provides sufficient volume to irrigate one hectare of corn. The total number of hours owned by each member is called a *tandeo* or *tanda*. The distribution of *acciones* among the members is the proportional allocation of permanent water rights as determined by the amount of the investment in the *galería* purchase or construction.

Once a *galería* system is operating, members can either sell their *tandas* for profit or use the water to cultivate their own land. They may also increase their water rights by buying *acciones* from others. Depending on the number of *acciones* owned, a *socio* can control a resource of considerable commercial value. Water, in the context of the association, has become private property and consequently is a marketable commodity. The ownership of shares (*acciones*) within an association changes over time and varies between associations: although some have been more or less stable, others have had a turnover in membership, or an internal reallocation of shares. We examine this process in greater detail later.

The irrigation water from each *galería* system is divided among the members in a thirty-day rotation or cycle of irrigation turns. The members receive water during specific time periods every month, depending on the amount of time to which they are entitled. The minimal unit for the measurement of water is one hour of the total flow of a *galería* system. Hence, there are 24 irrigation hours in a day and a night, and a month consists of thirty 24-hour irrigation periods, or 720 hours. Every month repeats the basic monthly turn schedule. The amount of time one farmer has in a month, the number of farmers entitled to water, the unequal length of the months, and the cyclical occurrence of leap years are incorporated into the system.[3]

The shares of the *galería* systems in the valley are based on three, six, or twelve hours of water, resulting in the establishment of an absolute number of shares available per system. A system based on three-hour shares has a total of 240 shares in a monthly cycle, plus 8 extra shares in each of the five thirty-one-day months and in a leap year; a six-hour-per-share association has 120 shares per month and 4 extra; a twelve-hour-per-share association has 60 shares per month and 2 extra.

In order to facilitate a comparison between *galería* systems, as a standard unit we have formulated the hectare-share (ha-s). The variable flow rates of the four *galería* systems that have resulted in the three-, six-, and twelve-hour irrigation periods per hectare can be compared by dividing the number of hours per member by the number of hours per month needed to irrigate one hectare of corn; the resulting value is the number of ha-s, which can then be added and subtracted as standard units of measure within and among *galería* systems. The total number of ha-s in an association also represents the irrigation capacity of a *galería* system.

There are two kinds of allocation: allocation or share distribution that represents the original investment in the associations (*acciones*), and the hourly distribution system within and outside the monthly cycles (*tandas*). It is important to keep the two concepts separate, because both *acciones* and *tandas* are bought and sold. The sale of an *acción* results in a permanent change in the ownership of a portion of the flow, whereas the sale or purchase of a *tanda* involves the use of a

Table 7. ***Water Flow and Hectare-Shares in Four*** **Galería** ***Systems***

Galería *System*	*Flow (Lps)*	*Hours per Hectare*	*Hectare-Shares in 30-Day Cycle*	*Hectare-Shares Not in Cycle**
Purísima No. 3	200	3	240	8
El Carmen	100	6	120	3
Guadalupana	75	12	60	2
Aldama	60	12	60	2
Total			480	15

Source: Enge 1982.

*On 29 February (leap year), 31 May, 31 July, 31 August, 31 October, and 31 December.

Lps: liters per second.

Table 8. ***Membership in Four Irrigation Associations and Average Hectare-Share Distribution, 1946 and 1973***

Association	*Hectare-Shares in 30-Day Cycle*	*Members in 1946*	*Hectare-Shares per Member*	*Members in 1973*	*Hectare-Shares per Member*
Purísima No. 3	240	85	2.8	75	3.2
El Carmen	120	52	2.3	52	2.3
Guadalupana	60	54	1.1	45	1.3
Aldama	60	52	1.2	43	1.4
Total	480	243		215	

Source: Enge 1982.

certain number of hours of water during a particular cycle, not a permanent change of water ownership.

As shown in Table 7, the maximum irrigation capacity of the four systems that serve Altepexi is 495 ha-s. Table 8 shows the distribution of shares among the members of the four irrigation associations between 1946 and 1973. Of the 511 participants in the ejido grant, 243, or 48 percent, became the founding members of the newly established irrigation associations; by 1973, 215, or 42 percent, were members.

In the four associations by 1973, a total of 27 *socios* had increased their shares, 36 had sold out, and 8 new *socios* had been admitted; the total membership had decreased by 28. Furthermore, in 1973, 23 of 215 *socios* (10.7 percent) were members of more than one association, and 19 of these had increased their holdings as the result of share transfers.

A total of seventy-five individuals were involved in share transfers of four different kinds: share increase, share decrease, sellout, and new membership. In the four associations between 1946 and 1973, twenty-seven people increased their shares, four decreased their shares, thirty-six sold out, and eight joined as new members. The forty who either reduced their holdings or sold out gave up a total of 93.6 ha-s, of which 25.4 ha-s were made available by reduction in holdings and 68.2 ha-s resulted from sellouts.

The recipients of the redistributed ha-s were the twenty-seven members who increased their ha-s by a total of 78.1 and eight new *socios* who acquired 16.2 ha-s. There is a discrepancy of 1.3 ha-s between the amount gained versus the amount sold, which is due to the use of shares not included in the monthly cycle. It is important to note that 83.4 percent of the total that changed hands went to the twenty-seven members who increased their holdings.

In order to show the relative distribution of shares held by the total membership in all four associations, we have classified members with three or more hectare-shares as *socios mayores*. Three or more hectare-shares provide a grain yield and resultant income that are well in excess of family needs, and these agriculturalists usually sell a significant portion of their harvest in the regional market in the city of Tehuacán.

In 1946, the *socios mayores* numbered 47 of 243 members (19.3 percent) and held 52 percent of all the ha-s in the four associations. By 1973, their number had decreased to 43 of 215 members (20 percent), but their ha-s had increased to 57 percent. In 1946 the five members with the most water rights held 89.6 ha-s, or 18 percent of all the irrigation water in the *municipio;* by 1973 their share had increased to 121.7 ha-s, or 25 percent. The remaining 172 members owned 197.1 ha-s, with an average of 1.15 ha-s per member.

The remarkable fact that emerges over the twenty-seven-year period is that there have been two principal sources of water shares for redistribution. The largest source was from thirty-six *socios* who sold out a total of 68.2 ha-s, and the other was the reduction by 23 ha-s in the holdings of a single *socio.* These two sources account for 91.2 ha-s, or 97 percent of the total amount of water that was redistributed. Furthermore, of the 91.2 ha-s, 55.1 went to the top five *socios.* A total of 17.2 ha-s were sold to the eight new *socios,* and the remaining 16.3 ha-s were distributed among the *socios mayores.* Sixty-four percent of the total amount of water that changed ownership went to the top five *socios,* and the principal source of the water was from the *socios* who were bought out. The ha-s held by the top five *socios* increased from 89.6 ha-s to 121.7, an increase of 32.1 ha-s. The change in the percentage of the total number of shares held by these *socios* went from 18.1 percent in 1945 to 24.6 percent in 1973. Of the 91.2 ha-s that changed hands during this time, the five top *socios* acquired 59 percent. The share distribution, which was never egalitarian, has clearly changed over the years, and a small group of *socios* have been able to increase their shares substantially.

The five owners of the largest number of shares stood out from the other *socios mayores* in 1973, not only because their holdings were in excess of ten ha-s each, but also because they were classified as the largest holders by other *socios* and informants in the *municipio*. Some referred to them as monopolizers of water and called them *acaparadores* (hoarders or black-marketeers), and in terms of holdings, they were the elite *socios*.

THE *SOCIOS:* RELATIVE WEALTH AND LIFESTYLES

We will now describe in some detail the five elite *socios*, their relative wealth, and how they differed from the rest of the *socios*. In order to place them in perspective within the *municipio*, we have included a brief description of Altepexi and its inhabitants in 1973.

The village of Altepexi was the principal population center of the *municipio* where the *socios* lived. The village consisted of a main square, municipal buildings, a Catholic church, a primary school, and a health center. On the northeast edge of town was a textile mill that produced a crude white cloth used for flour sacks. There were no other large commercial enterprises except approximately twenty-five family-owned stores and a gas station on the main highway between Tehuacán and Teotitlán del Camino. In 1973, all the streets were unpaved and most of the dwellings were made of adobe, with either sheet metal or tile roofs. There were eighteen houses made of brick and cement blocks. These were usually much larger and had metal doors, plate glass windows, and, in some cases, metal bars over the windows.

The inhabitants of the *municipio* were primarily agriculturalists, rural wage laborers (*peones*), mill workers, or wage laborers in the various factories and businesses in Tehuacán. The largest single occupation in the *municipio* was farming. Many families supplemented their incomes by making pottery or, more commonly, baskets. Basket making has traditionally been done on consignment to large wholesalers in Tehuacán, who provided the raw materials. The finished baskets were picked up by trucks or loaded directly into railroad cars for shipment to Mexico City and the United States (Enge 1977).

The population of the *municipio* was fairly homogeneous, and approximately 70 percent spoke a dialect of Nahuatl considered to be unique to the *municipio*. The Nahuatl spoken in Chilac and Ajalpan, though mutually intelligible, was easily distinguished and notably different.

The women in Altepexi wore blouses called *batas* that had a unique hand-embroidered pattern. Women wearing Altepexi *batas* were recognizable in marketplaces as far away as Córdoba in the neighboring state of Veracruz. Virtually all males wore modern clothing, except for the very old, who wore the traditional homemade white pants and shirt.

Informants in the *municipio* made clear distinctions of relative wealth and prestige within the local population. In a sample of household members, there was almost complete agreement that Julia Fraile, Juan Soler, José Arribas, Florencio Martínez, and Angel Santos (pseudonyms, as are other names used throughout) were the wealthiest and most influential individuals in the *municipio*. In addition to these five *socios*, José Martínez was mentioned. He owned the only gas station in the *municipio* and a small fleet of trucks, but was not an agriculturalist. All except one in the sample ranked Julia Fraile as the wealthiest person in the *municipio*. The other *socios mayores* were generally considered to be successful agriculturalists, but were not accorded the label of being rich or wealthy.

Julia Fraile lived with three of her five sons and two daughters in one of the largest cement block houses near the *municipio* offices (*ayuntamiento*). The two elder sons were married and had established their own households several blocks away. The sons' houses each had a medium-sized store in the front, which sold fabric, sewing notions, and patterns. In back of Julia's store was a large patio and, on one side, a modern kitchen with a new refrigerator and gas stove. Next to the stove was the traditional fire hearth and *comal* used for making tortillas and heating large clay vessels. Around the courtyard were five sleeping rooms, which were also used for grain storage, and one large room was used exclusively as a shrine for Julia's deceased husband. The shrine was empty except for a large altar at one end, lavishly decorated with pictures of saints, tinsel, and burning candles. In back of the first courtyard was another courtyard, approximately double the size of the first. It was bordered by high adobe walls and had a large gate at one end. Here were kept two late-model trucks, one about eight tons and the other a little smaller. In one corner was a tractor about four years old, and next to the tractor was a pen with two large bulls (*yunta*) used for plowing. The bulls were still kept because they were said to be more efficient at the plowing to shore up and support the corn plants. The tractor, with a lower clearance than the *yunta,* would damage the already tall plants. Julia also kept a large number of chickens and turkeys. In another corner was a pen with six pigs that were being saved for special occasions.

In addition to the two houses occupied by her sons, Julia also owned three houses in Altepexi and an apartment house with twelve units in Tehuacán. She went weekly to Tehuacán to collect the rent from her tenants, who were wage laborers in the hat factories and chicken farms near the city.

Besides her children, Julia's mother, who was nearly seventy-five years old, a younger brother, a cook, and three servants also lived in the house. Depending on the time of year, as many as twelve *peones* ate meals in the house and, during the harvest season, slept on straw mats (*petates*) in the courtyard. Julia hired the same *peones* from year to year, maintaining close personal relationships with many of them.

Whenever a *peón* was in need of medical attention or a small loan, he or she would come and ask Julia. She obliged virtually all "reasonable" requests. Several of her *peones* claimed that the Lord had blessed her with large amounts of money and water, which provided them with dependable employment.

There seems to have been little difference between the type of patron-client relationship that Julia had with her *peones* and that which existed under the hacienda system. The haciendas were much larger and employed many more *peones,* but the nature of the personal relationships was similar. The other elite *socios* also employed *peones* and had established similar relationships.

It is difficult to determine the absolute size of Julia's wealth in 1973, but considering the value of the trucks, tractor, and real estate, at a conservative estimate she must have been worth well in excess of 100,000 dollars. On subsequent visits to Altepexi during 1981, 1982, and 1983, we determined that her wealth had increased considerably—to double or even triple her worth in 1973, at a conservative estimate. Such wealth for a middle-aged Nahuatl whose family came from a line of *peones* at the Hacienda San Francisco is remarkable. Her wealth was in part inherited from her husband, but her own family, the Solers, had also become wealthy.

Julia Fraile's sons pursued different careers. Her oldest son, José, was a full-time agriculturalist who, for all practical purposes, was working for his mother, since he had no land or water shares in his own name. Another son, Miguel, married and moved out in 1971. He operated a transportation business using Julia's two trucks. Late in 1971 he had a serious accident with the largest truck. He was not hurt, but the truck was severely damaged. Julia promptly paid 300,000 pesos in repairs, and the vehicle was back on the road within three months.

Rodolfo, who was a year younger than Miguel, married in 1972 against his mother's wishes. He chose to marry a girl who was not originally from Altepexi and who was not the one Julia had chosen. He moved out of his mother's house several months before the marriage and went to live with his future wife and her mother in a run-down adobe on the outskirts of town. Six months later, they moved into a large house, owned by Julia, that was located about three blocks from where Julia lived. Rodolfo and his mother were not on speaking terms throughout this entire period, and he had to make a living without access to any of Julia's resources, except the house.

The important aspects of the relations between Julia and her three working-age sons were that José and Miguel married women who were the daughters of *socios mayores* and that they continued to work in agriculture and trucking. Trucking appears to have been Julia's favored investment of agricultural income. Trucks also provided a high rate of return when used for transport to local and regional markets. Rodolfo, on the other hand, went against his mother's wishes and married someone who had no potential inheritance. The woman Julia wanted him to marry was the youngest daughter of José Arribas, the third-largest owner of shares in the *municipio*.

Julia often said that she wanted her sons to continue as agriculturalists and that they should marry the daughters of other *socios* who had respectable holdings. She felt that this was the best form of financial security and that the family would always have enough irrigation water even if the flow in the *galería* system were to decrease markedly. She was, without doubt, the wealthiest and most cunning of all the agriculturalists in the *municipio*. Her personal wealth and plans to acquire more by strategic marriages clearly show her prominent position in the local hierarchy.

In 1973, Juan Soler, an old man well into his seventies who was no longer active in agriculture, was living in Tehuacán, where his sons were operating his trucking business. Soler had been instrumental in raising a large portion of the funds needed to purchase the P3 portion of the Purísima *galería* system. Where did Soler get the money to make the original investment of 60.6 ha-s (25 percent of the shares), worth 15,000 pesos in 1946, and why did he sell such a large number of shares during the years that followed? Many informants said he had found a buried treasure and made a wise investment when he became part of the P3 purchase. Treasure stories abound in all parts of Mexico, and they seem to be the most common way to explain why some have accumulated more wealth than others. It is a well-known fact that during the Revolution and the two decades that followed many people were engaged in smuggling, gun running, robbery, and other illegal activities. It is most likely that Juan Soler made money in some clandestine way and put it to use purchasing P3. Subsequently, he made a number of unwise business decisions that resulted in substantial losses. In order to raise capital, he gradually sold part of his irrigation shares to friends and relatives in Altepexi. In 1973, he was selling the *tandas* from his remaining 37.6 ha-s, and, with his sons, he continued to operate a trucking business.

The other three elite *socios,* Arribas, Florencio Martínez, and Santos, were well-to-do residents of Altepexi who lived in large cement block and brick houses. All three had invested in trucks and operated stores, located in their houses, where they sold a variety of goods ranging from fresh bread to agricultural tools. The stores were a kind of rural general store (*abarrotería*), which was also a bar and meeting place for the owner and his friends. These *abarroterías* also extended credit, and many of the regular customers had large outstanding debts, in some cases in excess of ten thousand pesos. Many of the debtors worked as agricultural wage laborers for the store owners. By providing easy credit for the *peones,* the elite *socios* could maintain a regular and dependable supply of labor, which was said to be becoming increasingly scarce.

The *socios mayores* with three or more ha-s of water lived more modestly than the elite *socios.* Some owned an old truck or a car, but none approached the lifestyle, residence quality, and material wealth of the five *socios* with the most shares. Nevertheless, the *socios mayores* were living at a level notably above the rest of the association members.

The members with fewer than three ha-s lived in smaller adobe structures and owned few of the valued consumer items such as refrigerators, record players, and television sets. Their houses consisted primarily of three or four rooms with a cement floor and relatively little furniture. Most of these members were self-sufficient in corn and beans, which provided them with a higher standard of living and nutrition than that of individuals with little land and no shares of their own. In order to supplement their agricultural incomes, some *socios* were basket makers (*canasteros*).

The preceding descriptions show that the *municipio* was stratified according to wealth as measured by the holding of shares in the irrigation associations. The informants clearly perceived the ownership of water as one of the principal sources of wealth, and the observed lifestyle and possessions of the elite *socios* bear this out. Four distinct groups of agriculturalists were evident: (1) the five elite *socios* who held the most ha-s; (2) *socios mayores* who owned three or more ha-s but fewer than ten; (3) the majority of the *socios*, who owned fewer than three ha-s; and (4) the agriculturalists, who held little land, owned no water, and had to buy *tandas*, sharecrop, or work as *peones*. As a consequence of water scarcity and high prices, many have abandoned agriculture and gone to work as wage laborers in Tehuacán.

THE DISTRIBUTION OF EJIDO LAND

As described in Chapter 4, the ejido grant made to Altepexi could either have been worked collectively by the eligible *ejidatarios* or have been divided into *parcelas* and cultivated by individual holders. The *ejidatarios* in Altepexi chose to divide the grant into *parcelas*, and one of their administrative responsibilities to the DAAC was to turn in a list showing the equitable distribution of land. When changes were made, they had to notify the DAAC immediately; consequently, there was constant communication between the DAAC and the local ejido officers. The first ejido commission complied with the requirements of the agrarian reform laws, and the participating *ejidatarios* received official certificates showing their legal ejido tenure of one *parcela*.

The DAAC did not have a permanent office in Tehuacán until 1976, when it was reorganized as the Secretaría de la Reforma Agraria (SRA). Before that, all administrative action and decisions were made in Puebla, and regional delegations and special representatives of the DAAC were stationed, for varying periods of time, in the Tehuacán area. Their purpose was to oversee the start of ejido operations and to resolve problems and disputes. Once they were established, the ejidos operated autonomously, but the government representatives were never very far away. Because land reform officials made unexpected stops in the *municipio*, the *ejidatarios* kept a current list of ejido landholders showing equitable *parcela* distribution, but actual land use was determined by the number of shares of water in the *galería* associations and by access to the small amount of ejido water.

As reported by several other investigators (DeWalt 1979; Hernández 1965; Romanucci-Ross 1973), the *ejidatarios* have been very secretive and sensitive about disclosing information indicating that land use is different from legal tenure. In Altepexi, many informants felt that an *ejidatario* could lose the right to participate in the land grant if any irregularities were documented. According to several influential agriculturalists in the *municipio*, the ejido commission did not keep any records showing an alternate distribution of *parcelas*, but in the process of establishing the distribution of irrigation water, it was possible to construct an accurate picture of land use in 1973.

Older informants, who were part of the original and many subsequent ejido commissions, said that land use in 1973 was more or less the way it had been for nearly twenty-five years. If this was indeed the case, then land use became stabilized after the irrigation associations were in operation. The members had been able to add to their own legal ejido holdings in proportion to their increase in shares. For 43 *socios mayores,* the amount of land held was equal to or less than the total number of ha-s. The difference between the amount of water and land is the ha-s that were available for sale or sharecropping (*siembra a medias*). It is also important to note that the hours of ejido water were distributed exclusively among the *socios mayores* in the *galería* associations. The rest of the ejido land in actual use was distributed among the 173 remaining members. Sixty sharecroppers were cultivating 541 hectares divided into 423 *parcelas* of the 1,638-hectare ejido grant. The area cultivated in 1973 corresponded very closely to the amount of irrigation water available: 541 hectares of land and 55 ha-s. The extra ha-s of water were probably sold to users outside the *municipio.*

If the official classification of *riego* (irrigation) land is applied to the area cultivated, twenty-eight members were using excessive amounts of land. And when using the criteria of 6 hectares of temporal land as a maximum, then seven socios were using land illegally. This amounted to 137 hectares and 75 hectares, respectively.

The amount of land in use during any one agricultural cycle varied considerably between 1945 and 1973. The area was always, however, directly proportional to the amount of available irrigation water. The principal reason for the variation was that in some years, large amounts of water were purchased by the *tanda* from systems outside the *municipio.* The maximum area cultivated in the *municipio* between 1945 and 1973 was approximately 750 hectares.

Siembra a medias is a contractual arrangement between a *socio* and an *ejidatario* who neither owns shares nor has the right to use ejido water. These arrangements occur when a *socio* chooses to use only a part or none of his or her water. Instead, the *socio* agrees to provide the *ejidatario* with all the necessary water and to pay half the cost of fertilizer and seed; in exchange, the *ejidatario* is to provide the land, all the labor, and half of the cost of fertilizer and seeds. The harvest is then shared equally.

For the most part, only the elite *socios* and the other *socios mayores* had sharecropping agreements in 1973, occasionally with agriculturalists outside the *municipio.* The five elite *socios,* Fraile, Soler, Arribas, Martínez, and Santos, had up to thirty sharecroppers (*medieros*). Once a *mediero* relationship was established, it usually continued for many years, and, in many cases, close personal relationships were established.

Because of hard labor and problematic supervision of *peones*, many *socios* decided that having *medieros* was better than cultivating themselves. During the five years prior to 1973, there was an increasing shortage of rural wage laborers, and several *socios* claimed that the *siembra a medias* was a very satisfactory alternative. Angel Santos went even further and said that, if ejido land tenure was ever enforced by the government and limited to the legal maximum, then he and all the other *socios* would get more *medieros* to make up for the lost land. At one point, he said the government could take all the land and he would still make money. In 1973, Juan Soler was the only *socio* who did not cultivate any land, but sold monthly *tandas* and had many *medieros*. In the past, he had cultivated some land, but never more than twenty hectares. In 1951, he became more involved in his trucking business, and every year after that he cultivated less land. By 1962, he cultivated none. He had always been strongly in favor of sharecropping, because it removed the constant worry about labor, and he felt that to be able to invest in water for a modest financial return with no worries about labor problems was well worth it.

The difference in actual income between *socios* doing their own cultivation and *medieros* was not very large in 1973. The cost of cultivating one hectare of corn with a yield of 1,400 kilograms was 940 pesos: 30 pesos for seed (15 kilograms per hectare), 600 pesos for labor (thirty person-days at 20 pesos per day), 220 pesos for fertilizer, and 90 pesos for the *yunta*. The corn was sold for 1,800 pesos per ton, which gave a gross income of 2,520 pesos with a net of 1,580 pesos. With a *mediero* providing all the labor and plowing and half of the fertilizer and seed costs, the net income for each was 1,135 pesos. The factor that made the difference, of course, was the labor cost absorbed by the *mediero* and his family.

Ejido landholders with no water shares of their own, who paid for the monthly *tandas*, would not have a monthly income equal to their costs during the dry winter months, between October and May. During the rainy season they could never predict whether there would be enough rain to eliminate purchasing all the required *tandas*. Consequently, virtually no one in the *municipio* ever risked cultivation without irrigation. Depending on how much water was purchased from outside the *municipio*, at least nine hundred hectares of the ejido grant has been unused and called *tierras ociosas*.

EJIDO AND *MUNICIPIO* LEADERSHIP

The elections for the formal ejido positions, as required by the Ley de la Reforma Agraria and supervised by the DAAC, take place every three years. Twelve positions have to be filled, but the only ones considered important are the president, secretary, and treasurer of the *comisariado ejidal*. To satisfy the government authorities, at least two groups or slates of candidates are put up for election. A recurrent problem is the lack of a quorum of *ejidatarios* on election day. The individuals who do not cultivate their *parcelas* see no need to attend the election meetings. To get a quorum, almost all of the active agriculturalists have to attend, and this is often very difficult to accomplish. Ejido representatives have been known to go from house to house tapping on doors in order to get a quorum on election day.

Between 1945 and 1973, the three most important positions were always filled by *socios mayores;* there were no exceptions. Out of these forty-seven individuals, twenty-eight had held at least one of the ejido positions, but no one had held the same position more than once.

The elite *socios,* with the exception of Julia Fraile, had all held the positions of president and secretary. Prior to his death in 1963, Ernesto Cofiño, Julia's deceased husband, had also held both these positions. The remaining eighteen positions were held at various times by other *socios mayores*. Whether or not they were actually holding a leadership position, the elite *socios* were known and constantly referred to as leaders by the use of terms such as *jefe* or *cacique*. It was understood that these five had complete authority in decisions about the distribution of *parcelas* and ejido water.

By now it is quite apparent that the elite members of the irrigation associations also controlled the distribution of ejido land in the *municipio*. They were also reputed to have made the important administrative decisions in both organizations. This leads to the question of the holding of *municipio* office and authority in local government in relation to status within the irrigation and ejido organizations.

Participation in *municipio* government is considered a duty by some and an inconvenience by many others. The most important *municipio* position, the *presidente municipal,* is a time-consuming job. It is either a routine position primarily concerned with settling minor disputes and officiating at functions and festivities, or, depending on who is president, a position with power and influence in the *municipio*. The presidents in Altepexi have ranged from illiterates who speak almost no Spanish to the wealthy members of the irrigation elite.

The *municipio* officeholders serve three-year terms, and their nomination by the PRI guarantees their election. In the period between 1945 and 1973, there were twelve *presidentes municipales,* two of whom were replacements for two presidents. Eight were *socios,* and of these eight, four were elite *socios:* Ernesto Cofiño, José Arribas, Florencio Martínez, and Angel Santos, who served from 1948 to 1950, 1954 to 1957, 1958 to 1960, and 1964 to 1966, respectively. Three of the remaining four were *socios mayores.* Arribas's term was particularly eventful because of the introduction of electricity from the Federal Electricity Commission. The *municipio* no longer had to rely on small, privately owned generators.

It is possible that the elite *socios* had considerable power and control over the *municipio* government, regardless of whether one of them was holding office. The political power center is not in the position of the *presidente municipal,* but in the local PRI party organization. The PRI provides one of the strongest and most direct links to higher levels in the Mexican governmental hierarchy. It is, at times, difficult to separate the PRI from the various government agencies. That is to say, the dominance of the single political party has made the holding of the government position and party membership synonymous. The resultant patronage system is apparent at the *municipio* level as well as in the higher levels of the bureaucracy.[4]

The elite *socios* and some *socios mayores,* the reputed political leaders of the *municipio,* name the candidates for the *municipio* offices and are occasionally candidates themselves. When these *socios* hold *municipio* offices, they consistently demonstrate their influence by implementing such projects as electrification, the introduction of potable water, and school construction. For a small *municipio* to become the beneficiary of large-scale government development projects, extensive political manipulation within the hierarchy is usually required. The PRI leaders have the connections and are instrumental in determining which federal projects will be implemented.

The ability of a political leader to obtain outside funding and technical assistance for projects to benefit the *municipio* is only part of the total picture. Virtually all local and relatively small-scale development projects in Mexico receive only partial funding, with the balance paid by the recipient population. Consequently, local leaders must also have the ability to organize and collect payments from the local population. They must convince residents that a costly project is both beneficial and necessary, and this, according to some, requires considerable persuasion.

The local leaders of the PRI, in return for favors from the top, guarantee that the officially sanctioned candidates are elected and also provide other services to the PRI party machines. One common request from the regional PRI office in Tehuacán was for the president in Altepexi to provide up to five hundred people to participate in parades and rallies in support of candidates for higher state offices. Other responsibilities of the local leaders included giving frequent fiestas for visiting PRI officials from Tehuacán, Puebla, or Mexico City.

STATE CONTROL OVER LOCAL IRRIGATION ORGANIZATIONS

In the course of studying local irrigation organizations, we became aware of similarities in Altepexi, Ajalpan, and Chilac. The decentralized irrigation associations all had comparable bylaws, personnel, election procedures, and operation modes. Only by comparing our data and looking at the role of state institutions were we able to understand the mandate of the SRH, as formulated in the legal statutes of the agrarian reform, to clearly define the organization, rights, and obligations of groups that were granted permission to use water resources (Ley Federal de Aguas 1936 and 1972). An examination of the relevant statutes is most illuminating for identifying the rules that are at times obeyed, bent, or ignored, but that over time have shaped and given distinctive characteristics to both structure and behavior.

The Mexico water laws define the procedures that must be used to petition for access to federal water, how to enforce the conditions and regulations that go with access and use once permission has been granted, and the organization and conduct of associations formed for the purpose of using water. Our interest is primarily in the last, but an examination of the entire hierarchy of procedures and administrative regulations shows how all aspects connect.

The SARH has the power and duty to determine the rights of potential users, to avoid wasting scarce water, and to distribute water for "maximum benefits." It determines the flow rates per second or volume per year for each user and also sets the time and manner of use. Technical studies, needs of local users, relevant documentation, and meetings with potential or traditional users are part of the ministry's procedures for making permanent and temporary concessions. The individuals or groups interested in gaining concessions must form a *junta*, or assembly, under the direction of the SARH. Records must be kept of all activities, and suggestions for granting rights and stipulating the conditions for use are subsequently taken into consideration by the ministry.

The conditions and regulations for water use consider the cost of distribution, the location of land to be irrigated, existing infrastructure, existing use patterns, dispositions of petitions for use, and the ability of current and potential users to contribute to the upkeep of existing and proposed infrastructure. Furthermore, the users must form an association, or *junta de aguas,* to assure compliance with the conditions set forth by the SARH.

Once a concession has been made and a permit given, the *junta de aguas* must hold a general meeting with one or more legal representatives of each group of users. The users can be ejidos, national irrigation systems, *municipios,* associations of users, irrigators owning private property, or industrial companies. If, however, there is only one group of users, one person will be named as water judge, or *juez de aguas,* and will assume the responsibility for complying with the conditions and following the rules for water use. In their role as enforcers, the *junta de aguas* and *juez* function as agents of the ministry and have clearly stipulated rights and obligations.[5] When there is an indication of infractions, the *junta* or *juez* can call on the ministry or appropriate authorities to investigate and rectify the situation. Furthermore, the ministry can authorize the *junta* or *juez* to cut off the water to users who are behind in their payments. In all cases, detailed reports of all actions must be submitted to the ministry. The ministry can inspect the conduct and work of the *junta de aguas,* and if something is amiss, a special assembly of the users will be convened to name substitutes (Ley Federal de Aguas, Ch. 20).

In communities sharing irrigation water from rivers and streams with many groups of users, the *junta de aguas* is a prominent part of water distribution and political organization (Downing 1974; Hunt and Hunt 1974). In the case of the *galerías* in the Tehuacán Valley, single groups of users have been given permission to construct independent irrigation systems. Here, instead of a *junta de aguas* in each *municipio,* associations of users of each *galería* have been formed.

An individual who wishes to organize an association of water users must have a document stating the purpose, the age and address of each person involved, the name and location of the water resource, the hectarage of the area to be irrigated, a description of the water works, value of the works, and the capital investment of each potential member. Once the association has been given a legal charter by the ministry, the members are told how to elect a *junta* or *mesa directiva* and *vigilancia* (oversight); the term of each *junta;* causes for *junta* removal and substitution; rights and obligations of the *junta;* when routine and emergency sessions should be held; the rights and obligations of the president, secretary, voters in an assembly (*vocales*), and treasurer; and the rights and obligations of the members.[6]

The *junta directiva* is the legal representative of the association and can deal directly or through counsel with the authorities. It can also make contracts and agreements with individuals and groups (i.e., other associations and organizations) and is responsible for administering the resources of the association, carrying out resolutions made by an assembly of the members, hiring and firing employees, and applying sanctions to members who break regulations. Specific obligations of the *junta* include the convening of assemblies, formulation of budgets and plans, reporting on working progress and planned projects, and making financial reports. The format for the conduct of assemblies is also given, and the organizational format, as described earlier in this chapter, is very much in accordance with the rules. The associations are obliged to keep minutes of meetings and accurate records. The transfer of land or water within the jurisdiction of the association brings the new owner into the association, and he or she must apply to the *junta directiva* for membership. The previous owner stops being a member when he or she has no more land or water. If an owner has not approached the association for membership, the *junta* must send a letter asking him or her to join. If the owner does not respond, the *junta* can ask the federal court for the right to expropriate both land and water, subdividing them among the members (Ley Federal de Aguas 1936, Chap. 21).

Both ejido and irrigation associations have structural and procedural similarities that are imposed by government regulations. Individual landholders and water owners work within this framework and, depending on a person's status, role, and connections, they have varying degrees of latitude. Specific circumstances, people, and events determine the kinds of action, both legal and illegal, that can take place.

SUMMARY

Agrarian development of the Tehuacán Valley has been based on expanding the supply of irrigation water. The state has not played an active role in the process. Instead, the indigenous population has, in a variety of ways, mobilized capital to develop water resources. The most common form of organization has been the *galería* association.

Galería associations have generally included people from different economic strata, but have been dominated by the most wealthy families in the valley. In the cases where *galerías* have developed large flows of water, the members of the associations have made considerable profits on their investments.

In Mexico private ownership of water runs counter to law, but it is allowed in some regions where there is little or no state investment in irrigation infrastructure and the indigenous population has allocated large amounts of capital or labor to develop water resources. How the water is distributed becomes a critical issue, one that we examine in the next chapter. Under most systems, unequal control over critical means of production generates greater social stratification and class formation within communities or regions, unless there are internal mechanisms to redistribute the accumulated wealth. The concentration of control over the means of production has the potential to generate antagonism and conflict between different segments of the population.

The irrigation associations were organized to increase the irrigation water available to groups of ejido landholders. This was achieved by the combined financial resources and labor of distinct subsets of the total ejido population, which was most probably organized and structured according to an existing hierarchy of wealth, influence, and control. Another limiting factor for group size was the volume from any one *galería* system, which, depending on the actual flow rate, could only provide water for fifty to one hundred individual *parcela* holders.

The allocation of water rights among the association members was and is based on the original investment and the subsequent purchase or inheritance of additional shares. The distribution of shares is reflected in the monthly cycle of *tandas*. Both permanent water rights and individual turns are marketable commodities at the disposal of the members.

In the case described in this chapter, the distribution of shares in the four associations was neither egalitarian nor monopolized by a few members. Some were able to increase their holdings over a twenty-eight-year period, but no major redistribution occurred. The data show that a number of *socios mayores* and elite *socios* substantially increased their holdings, and an economic elite based on wealth from water was established in the *municipio*. At the same time, the associations provided a reliable source of water for a majority of agriculturalists who raised crops primarily for their own family needs. In this case there appears to have been no overt pressure for smallholders to sellout to *socios mayores*.

There were three distinct classes of *socios* based on relative wealth, lifestyle, and classification by informants in the *municipio*. The ownership of shares was clearly a means to acquire wealth that was used by elite *socios* to invest primarily in trucking. Wealth based on the amount of water that was owned differentiated ordinary *socios* from the *socios mayores* and the five elite *socios* (Fraile, Soler, Arribas, Martínez, and Santos).

The organizational structure and bylaws defined the positions that were to be filled and the rules that were to be followed for administering an association. There appeared to be no correlation between the frequency with which a member held office and the size of his or her ha-s—that is, the *socios mayores* did not hold office more often than the other *socios*. Since most positions were routine, it is clear that officeholding in general was not a good indicator of internal association hierarchy. However, when one categorized the types of positions held, the *socios mayores* clearly dominated the crisis commissions.

Consequently, the role structure must be analyzed and separated into distinct groups according to function in terms of routine operation versus unusual or crisis situations. This is particularly important if we are to identify the central role or roles that provide unification within the organizational structure.

Ejido land tenure was clearly subordinated to the ownership of shares or a portion of the ejido water from La Taza. In other words, if an *ejidatario* had enough irrigation water, there was no effective limit on the amount of land he or she was able to cultivate. The *siembra a medias* was an important mechanism whereby the water owner did not have to physically cultivate the land, but was still able to earn a substantial income from water shares.

The elite *socios* and the *socios mayores* were an integral part of the ejido administration, regular office-holders in the *municipio,* and leaders in the *municipio* organization of the PRI. Clearly then, individuals who have substantial holdings in the water associations have also been involved in other key aspects of ejido and *municipio* administration. Informants have identified the elite *socios* as caciques, which indicates that they had power and influence not only within the irrigation associations, but in their society.

Even with a differentiation of specific roles and their functions, it is difficult to describe the internal relations of administration and control in the irrigation associations. Structure, in a sense, helps to define specific roles, but only by studying real people and real events, over a period of time, can we understand how decisions are made. The next chapter examines the irrigation associations in the context of a variety of different situations. From the outcomes of specific circumstances and action, it is possible to analyze the centralized role structure, the amount of influence, and the extent of the control exercised by the irrigation associations and their personnel.

The exact nature of government intervention in local affairs can only be determined by examining events. It is clear that within the structural relations between local, regional, and national levels of organization there are definite legitimized channels for government intervention at any level. We will show, using documented events, the nature of state versus local control of productive resources in the Tehuacán Valley.

6.

ELITES AND IRRIGATION ASSOCIATION MANAGEMENT

THE STRUCTURAL relations of water control, regulation, and distribution are sterile without an examination of actual events to illustrate how power is, in fact, distributed and used. We have described differential water distribution and divided the *socios* according to a three-level wealth hierarchy; we now examine how the elite have behaved in the management of specific conflicts.

The events described here have been selected to show the range of activities controlled by the irrigation associations and those that involve, to a greater or lesser extent, outside individuals and institutions. The events are representative of the following types of association activities: procedures for the selection of leaders; rules for the sale of water and sharecropping; disputes over share ownership; collection of overdue payments from the members; charitable contributions to individuals, the local *municipio,* and the church; personnel problems—the water guard; construction and maintenance projects; and conflicts with other irrigation associations.

The documented association activities were the result of initiatives by particular members and supported by the others. Pinpointing individuals who continually take important initiatives serves to identify the leaders, not in terms of officeholding, but as individuals who have made instrumental decisions, that is, the key actors in the centralized role-set of the corporate group.

The leaders and their relative domains are viewed in terms of action within the irrigation association, but the extent of their domain can best be measured in their relations with outside institutions and individuals. Secure leadership and control within a local group can be pervasive, if not absolute, but as activities extend out into the *munici-*

pio and beyond, power begins to fade. Since conflicts and normal activities transcend these boundaries, documentation of events that cross multiple boundaries will tell much about the nature of local power brokers, their organization and mode of operation.

Over the past forty years, problems and conflicts between irrigation associations in the valley have been common. Purísima No. 3 (P3) has a long history of disputes with another association, Purísima No. 1 (P1). The basis of their constant bickering is the forced sharing of water from a single *galería* system; the resulting relationship has created tensions not observed in the operation of other associations. The particular history of P1 and P3 demonstrates the extension of control and influence of each association and the role played by various agencies and ministries of the Mexican government that have acted as arbitrator and enforcer of regulations and restrictions.

The nature and complexity of operations, management, and conflict have varied from association to association, but the type of problems and operational prerequisites are common to all who irrigate with *galería* technology in this region of Mexico. The events we describe occurred between 1945 and 1983 and have been selected from the records of Purísima No. 3, El Carmen, and the recollections of association members.

ASSOCIATION LEADERSHIP

To determine whether certain *socios* made more decisions than others and whether a relationship exists between the number of shares held and participation in formal leadership selection and how the *socios* reach consensus on projects and activities, the operation of associations must be examined and analyzed over time. In the process of reaching a consensus, members often passively agree, violently disagree, argue, debate, and shout, but eventually decisions are made. One shareholder's will, desire, or power in the form of a strategy or plan of action to confront a specific problem or crisis is eventually accepted by the rest.

Between 1946 and 1969 association officers and special commission members in Purísima No. 3 were selected from a small portion of the entire membership, and these members rotated holding elective office. The apparent reason for limited participation was made clear in a meeting of the P3 membership held in December 1956. At that meeting, José Arribas said there were many members who contributed neither time nor effort to association affairs. He proposed that more members be urged to take part. Another member suggested that the nominees be selected in order of appearance on the *lista de tandeo*. Angel Santos, as spokesman for the *socios mayores*, responded by saying that in case of special commissions for annual festivities this could be done, but for important positions to govern the association it would not be possible. In the governing positions, he continued, the association needed people who were literate and who could represent the association in all water-related matters. The membership agreed to continue to select nominees from the same group of "experienced" members. The elite members had in effect convinced the rest that only they should hold positions of responsibility; this certainly accounts for the high frequency of elite *socios* and *socios mayores* who have served on crisis commissions.

This exclusive selection process continued until December 1969, when Florencio Martínez, the president, and Angel Santos proposed that candidates be selected according to the *lista de tandeo*. Pedro Bartolomé lent his support and the proposal was approved by the membership. From 1970 until the end of 1973 the selection of nominees was taken from the *lista de tandeo* so that every member was obliged to take a turn and serve as an association officer; the crisis commissions, however, continued under elite control. After 1970, members could be sanctioned for not serving when their names came up on the *lista de tandeo*, as illustrated by the following example.

In 1972, Javier Valle was elected secretary of P3, but he did not attend any meetings during January and February. Consequently, the president named another shareholder to serve in his place. Pedro Bartolomé was chosen because his experience and literacy were considered essential to help resolve the problems with P1. The following year, Pedro Bartolomé claimed that Javier Valle should have reimbursed him for his time and asked the association to take care of the matter. Carmelo Sanfrutos suggested that Valle should be made to pay Bartolomé 500 pesos; Angel Santos proposed that it should be 1,500 pesos. The membership approved Santos's proposal and Valle was notified by letter of the association's decision. Valle did not pay. In January 1974 Angel Santos and José Arribas proposed that the association would sell Valle's *tandas* until the fine was paid in full. At that point Javier Valle agreed to pay 30 pesos a week until his debt was paid.

Again, Santos and Arribas, two elite *socios*, determined the outcome and amount of money owed for noncompliance. In this case, noncompliance was the refusal to do administrative work for the association as prescribed in the bylaws and reinforced by the resolution that all members must serve when it was their turn.

THE SALE OF WATER AND SHARECROPPING

The following account details an association prohibition on the sale of water to nonmembers and the termination of sharecropping agreements. The case demonstrates elite *socio* control regarding the disposition of *tandas* and shares.

In the early part of 1954 there was extensive unrest, agitation, and protest over the lack of water on the part of agriculturalists in San Diego Chalma, San Pablo Tepetzingo, and Altepexi. The protestors complained that the canals with large volumes of water flowing alongside their dry *parcelas* were being unjustly monopolized by the members of the irrigation associations. Furthermore, they claimed that the irrigators were putting too much water on their land, that the same volume could be used to irrigate more land. During this time rainfall was far below normal in the whole region, and extensive drought drastically decreased crop yields.

It has not been possible to reconstruct how widespread and militant the protest movements were. There are no written accounts, nor does anyone recall any unusual theft of water at that time, but the situation was serious enough to prompt a rapid and strong response by the concerned members of P3. Some members felt threatened by the protests, fearing theft and violence. At the time, approximately ten P3 shareholders were selling *tandas* and had *medieros*. Many members felt that this was a major factor contributing to the unrest. The high prices charged for *tandas* because of the drought must have been an important variable, and continued sales or *mediero* contracts could only have aggravated a tense situation.

In March 1954 the association held an emergency session to decide on what action was necessary to protect the association's water. At that meeting, the president of P3, Martín Iglesias, asked the members to be very careful to avoid provoking any incidents. The membership had a heated discussion as to precisely what should be done to protect the association's interests. Florencio Meléndez proposed that in order to protect the interests of the association and to prevent possible violent disturbances, the association should stop the water flow in the canals passing through the problem areas. Furthermore, he suggested that the corn planted by *medieros* in these areas should be harvested as soon as possible, that there be no new water contracts, and that sharecropping be stopped. Angel Santos supported Meléndez and asked the assembly to calm down. He said that they should not make a resolution without first carefully investigating what was going on and proposed that each member who had contracted water for money or crops should explain his or her individual case and indicate any possible problems. This, he felt, was the best way to find the source of the agitation; bothersome problems like this were costly in both time and money. The assembly agreed that in the next week's session they would decide what to do based on more accurate knowledge of the situation.

In the next session, at the urging of Angel Santos and Romualdo Madariaga, the membership voted to suspend all sales and sharecropping contracts between association members and outside agriculturalists. The crops under contract should be harvested early to prevent loss in case the situation got worse. The main concern was with the corn owed by *medieros* for water already used. Florencio Meléndez insisted that a severe fine be levied on any member who did not comply. Santos and Madariaga agreed, proposing a charge of two hundred pesos for every twelve hours of water sold for money or crops to any nonmember. The prohibition on the sale and rental of water became known as the Decree of April 1954 (Decreto de abril de 1954).

Approximately one year later, the water guard (*aguador*) informed Martín Iglesias that Paloma Plaza had sold four hours of water to someone in the neighboring *municipio* of Chilac. Iglesias claimed that Plaza had first sold nine hours and later another four. At a meeting in August 1955, several members asked that the Decree of April 1954 be enforced and that Plaza be fined at least two hundred pesos. Paloma Plaza defended herself, asking why only she had been fined, since both Florencio Meléndez and Romualdo Madariaga had also sold water. She claimed that Meléndez had sold twenty-four hours of water on May 16, 1955. Meléndez asked for proof, since he had only twelve hours of water during the night of the sixteenth, and he asked Antonio Molina to confirm that he had lent him those twelve hours. The membership agreed that Plaza's accusation could not be proved. Madariaga said that the last time he had contracted his water for sharecropping was in January 1954, before the decree went into effect, and that he and Juan Soler had been granted permission by the association. After receiving their grain payment in the early harvest they did not renew the contracts. José Arribas asked the president to authorize an investigation, but nothing was ever reported to the membership and no fines were imposed. Paloma Plaza eventually paid the fine when the association put pressure on her and threatened to sell her hours.

It is probable that many members were illegally selling water at a large profit. Plaza was singled out because she was not part of the leadership and could be an example for the rest of the members. Both

Meléndez and Madariaga were in all probability also guilty, but their positions as *socios mayores* and their support from the membership prevented any challenge to their versions of the story that could have resulted in formal accusations.

A DISPUTE OVER SHARE OWNERSHIP

Disputes over share ownership, transfer, and inheritance have been rare. According to both records examined and interviews with *socios*, the case described below has been the only one. This dispute was particularly interesting because the Ministerio Público became involved in the arbitration. Normally the ministry is involved only in cases in which statutes are clearly broken, and it will then prosecute in the tribunals in Tehuacán. The reasoning is that an individual who is using water not legally his or hers should be prosecuted for theft. The only problem is that theft can be proven only once ownership of the shares has been resolved.

In July 1970 the association received a letter from Catalino Ruiz, who asked to become a new member after having purchased a number of shares; the exact number and the proper documentation were not included. As a result, the president of the association rejected the request, calling for more details and a legal copy of the transfer of water shares. In August Ruiz sent another letter, which was also rejected by the association. According to the *aguador*, however, Ruiz had already started using the water. In January 1971 the association received a letter from the Ministerio Público in Tehuacán asking the association to recognize Ruiz as a member. The president said that Ruiz had been recognized, but he still needed to produce the proper documents so that the transfer of shares could be properly recorded. Three months later the association received a letter from Alfonso and Elpidio López accusing Catalino Ruiz of having falsified the signature of their deceased sister, María Isabel López, the former owner of the shares and an association member. The López brothers asked the association not to recognize Ruiz's right to ownership without first examining the proper documents. Claiming to be the rightful heirs of their sister's shares, they felt that they should be admitted to the association as new members. At a meeting in April 1971, Arturo San Martín said that neither the López brothers nor Ruiz had been favored by the association and that the matter was now in the hands of the authorities, who would decide the rightful ownership of the shares. Catalino Ruiz was, however, allowed to continue using the water.

The association received another letter from the López brothers on September 30, 1971, again accusing Catalino Ruiz of having falsified the signatures on the sale of documents. The letter claimed that Maria Isabel López had left the shares to her brothers, as documented by the justice of the peace in Ajalpan. The association still refused to take sides pending resolution by the authorities. In October, however, the association recognized and duly recorded in the association books the documents presented by Catalino Ruiz.

Ruiz successfully acquired the twelve hours of water originally owned by Isidro Casteñeda and inherited by his widow, María Isabel López. According to senior association members, Catalino Ruiz had indeed paid for the shares and was legally entitled to them. Further examination showed that Ruiz was a cousin of Martín Iglesias, a *socio mayor* who had actively supported Ruiz in his effort to become accepted by the association. The role played by the Ministerio Público was only secondary, because theft could not have been proven, and the documents from the justice of the peace in Ajalpan were never produced. In reality, the association was free to accept as the rightful owner whomever the leadership chose. The fact that Ruiz was permitted to use the water confirmed that the elite *socios* had given their tacit approval by allowing the change in the *lista de tandeo.*

ASSOCIATION COLLECTION OF OVERDUE PAYMENTS

The smooth operation of irrigation associations depends on prompt and regular payments for operating expenses. An interruption in the cash flow could seriously affect projects such as canal maintenance and *galería* cleaning. Frequently, the association has to exert pressure on members delinquent in their regular monthly payments and special assessments. The collection procedure involves three stages. First, the member gets a verbal reminder from one of the association officers. If this does not result in payment, a fine is levied. The fine is a percentage of the total owed, depending on how badly the association needs funds, ranging from 20 to 50 percent. If the member still does not pay, the association sends a letter demanding payment by a certain date. If the payment is not received by the deadline and no arrangement has been made for a partial payment, the association begins selling the member's *tandas* until the debt has been paid in full.

There are few documented cases where water has actually been sold, but the associations usually resort to letters and deadline setting. There are no known cases of any member who neglected payments to the association and continued to receive irrigation water. The threat is real, and the loss of a single *tanda* could result in a drastically reduced harvest or none at all.

ASSOCIATION CONTRIBUTIONS TO THE *MUNICIPIO*, CHURCH, AND INDIVIDUALS

The irrigation associations are viewed by other institutions and individuals in the *municipio* as very wealthy and a potential source of money in times of need. Depending on the circumstances, the associations have made many contributions. Although donations are, at times, politically motivated and repaid by favors at a later date, the associations do not function as banks or moneylenders. They are often approached by the *municipio* authorities and the church for contributions to building projects and occasionally to help in cases of individual hardship.

In December 1957, P3 received a letter from the president of Altepexi asking for assistance to pay for the lights in the central park in front of the *municipio* offices. The association voted not to contribute because of the extensive and costly canal cleaning that was still pending. Furthermore, the members felt that a recent contribution to help pay for school construction had been sufficient.

Another letter was received in June 1961 from the Commission for Church Construction and Remodeling, requesting contributions to reconstruct the priest's living quarters, said to be on the verge of collapse. Fernando Calatrava proposed a contribution of two pesos for each hour owned, another member suggested one peso. The latter was approved by the membership, because helping the church was considered the duty of everyone in the *municipio*.

The Commission for Church Construction and Remodeling sent another letter during April 1972. This time, the church specifically asked for 2,000 pesos, but the association contributed only 1,460 pesos. The contribution consisted of two 1-peso-per-hour charges per member, to be paid in weekly installments.

In July 1963 the *socio* Alfredo Simón appealed to the association for money because he was very sick. He reminded the association that on many occasions he had contributed labor to clean the canals. After much discussion as to whether or not the association should make a contribution, in the end they decided to turn down the request, reasoning that the treasury was very short of money because a *galería* cleaning project was taking longer than expected.

When a contribution was to be made, the members were required to pay immediately, which indicates that there was never much money on hand. Additional charges per hour could cause serious financial burdens for some members. If many extra charges for whatever reason had already been made, the members were reluctant to make charitable contributions.

THE ASSOCIATION WATER GUARD

In a small *galería* system, such as El Carmen, Guadalupana, and Aldama, the job of supervising and guarding the water allocation and distribution network is usually done by *socios*. P3, because of its large size, has a permanent full-time employee, the *aguador,* whose position is one of trust and importance to the association. The *aguador* is responsible for overseeing the flow of water in the canals, the opening and closing of gates at the proper time, and the prevention of water theft and loss due to spillage. By local standards the *aguador* gets a good wage, a house, and water for his fields. Problems with the water guard are always serious and result in immediate action by the association. A water guard who cannot be trusted cannot be tolerated by any association.

The case of Pedro Hernández, who was the P3 *aguador* between 1945 and 1951, is a good illustration of how this key person could cause problems in the distribution of water. Hernández lived in a house provided by the association. It was located along the P3 canal near the Ladrón de la Huerta. He was frequently accused of causing major water spills (*mermas*), to which he responded by blaming the poor conditions of many canals that needed cleaning and repair.

At a meeting held in February 1951, Gregorio Ortega proposed that Hernández be replaced, because the members could no longer tolerate the constant *mermas*. The membership was in general agreement and claimed that the losses had drastically decreased the amount of water during their *tandas;* if such carelessness were allowed to continue, their yields would be seriously affected. Consequently, the proposal to get a new *aguador* was unanimously approved.

At an association meeting in March 1951, Gregorio Ortega wanted to know why Pedro Hernández still had not been replaced. Hernández defended himself by saying that Ortega should complain to the owners of La Huerta, a ranch in the area the canal passed through, because they, not he, were responsible for the *mermas*. Furthermore, he said that he would personally compensate Ortega for four hours of water if he did not get any compensation from the owners of La Huerta.

There are no records showing that Hernández tried to compensate for any water losses. Instead of seeking damages from Hernández, the association continued to look for a new *aguador*. In March 1951 Angel Santos informed the assembled members that the association had asked Hernández to vacate the house belonging to the association. Santos said that the conduct of the *aguador* had been intolerable and that the association had proof that he was guilty of the following infractions: neglecting and abandoning his duties, changing water measurements in favor of another association, and working as *aguador* for another association, thus abusing the confidence of P3. Santos demanded that he be removed immediately.

Other members made further accusations against Pedro Hernández. Gregorio Ortega said the water he had lost went to the land of Hernández's friend, which, in no uncertain terms, amounted to theft. José Castello said he accidentally met Hernández one day and found him working for La Huerta instead of doing his duties for P3. Pedro Jordán said that in February, Hernández had a drunken orgy with many of his friends at a house in Pantzingo, where he was heard to have said he did not give a damn about P3 and its water. Florencio Meléndez, Santiago Herrera, and Alejandro Arce claimed that Hernández did not divide and distribute the water the way he should. When they were scheduled to receive their turns, they found obstacles intentionally placed at the water gates, and Hernández was found asleep in his house. The members were convinced that Hernández could no longer be loyal to P3. According to his own sons, he was more interested in working for another association.

After the assembly voted unanimously to remove Hernández, they made two resolutions: (1) Because Pedro Hernández was also working for another association and was illegally diverting P3 water, he could no longer work for P3. Both allegations had been proven. (2) In accordance with Chapter 5, Article 42 of the Labor Laws of Mexico (Código de Trabajo), noncompliance with specified duties, being drunk, and behaving improperly while on the job were just causes for immediate dismissal. Pedro Hernández moved out the next day.

The dismissal procedure and actual removal of the *aguador* was a very serious matter, and the *socios* were careful in documenting and proving the water loss and misconduct charges. Angel Santos took charge of the proceedings and made sure that all the legal requirements for discharge had been met. Compliance with the articles of the Labor Laws was necessary in order to avoid countercharges by Hernández. A judgment in favor of Hernández against P3 could have been costly in time, money, and further water losses.

This case demonstrates that the *aguador*, trusted by P3, was bought off by other interests and was able to steal water with impunity. It seems strange that the affected members did not protest sooner, unless they were involved in some way. Hernández could have been part of a larger scheme to divert a portion of the P3 water. Ever since this occurrence, the *aguador* has been under very close supervision. In 1962 the *aguador* was dismissed after committing only minor infractions. He was found guilty of neglect and failure to report small damages to the main canal.

CONSTRUCTION AND MAINTENANCE ACTIVITIES, 1945–1973

Since the paramount function of the irrigation association is to maintain a steady flow of water in the system and to oversee the orderly distribution of *tandas*, there are always ongoing projects involving canal cleaning, *galería* deepening and extension, and the repair of sluice gates. New construction of canals and *galería* extensions occur at greater intervals to maintain and increase flow rates.

The *socios* are constantly preoccupied with the water flow to the point of obsession. The loss of a single *tanda* can mean near ruin or a drastic reduction in the yield per hectare. Water losses commonly result from breaks in canal walls or excessive spillage leading to an interruption in the flow. By the time the flow is restarted, a *socios*'s water cannot be recouped without taking time from the next *socio*'s *tanda*.

Association meetings provide a forum for deciding on work that has to be done on particular parts of the system. The governing officers and special commissions hire the work crews, supervise the day-to-day work on the association projects, and discuss problems. The most common problem is a shortage of funds.

The financing of all association projects is accomplished by the collection of regular monthly payments based on the number of hours of water owned by each member. The collection of extra monthly payments when needed, fines, payments from new members, payments for share transfer, and income from the use of water by hydroelectric plants provide additional funds. Routine canal cleaning and miscellaneous association expenditures are paid from the regular monthly assessments and other routine income. Major construction projects are beyond the means of regular association income and must depend on additional payments from each member.

The regular monthly payments vary in amount between associations. In Altepexi two small associations, Aldama and Guadalupana, charged 1 peso per hour of water, and additional charges were infrequent between 1945 and 1973. A charge of 1 peso per hour resulted in an income of 730 pesos per month, and extra charges were usually equal to or multiples of this amount; that is, a 3-peso-per-hour extra charge gave an income of 2,190 pesos. Because of the relatively infrequent *galería* cleaning projects and the more frequent but cheaper canal maintenance, the shareholders were usually not charged more than three additional times during any one calendar year. In the case of El Carmen, continued *galería* construction in addition to canal maintenance resulted in a higher monthly charge of 2 pesos per hour. Additional charges were frequent in order to cover constant cost overruns. Because of the long *galerías* and large canal network, P3, the largest association, had by far the highest operating costs, but the regular monthly payments were maintained at only 2 pesos per hour. Additional funds were raised by charging the membership additional fees at frequent intervals. These charges have been as high as 5 pesos per hour per week for up to twenty-six consecutive weeks. The circumstances and details of costly projects paid for by large and frequent extra charges are described below.

The *galería* systems in the Tehuacán Valley have canals that go over and *galerías* that go under the land of *municipios* and ejidos that do not use the water. This has created a situation in which irrigation associations have negotiated with the *municipios* and ejido commissions for *galería* and canal permits. Ejido commissions are within their rights to deny permission for canals that could damage existing infrastructures and *parcela* divisions. Furthermore, the irrigation associations have to get permission from the SARH officials in Tehuacán to construct *galerías* and to enter existing ones for the purpose of cleaning and deepening. The access to water in the form of excavation permits and the right-of-way for the passage of canals have resulted in the formation of a complex web of relationships between the associations, the SARH, *municipio* authorities, groups of landholders, and various ejido commissions.

Depending on the individuals and organizations involved, the negotiations for the right-of-way to construct under or on the ground can be both lengthy and costly. When an association tries to reach an agreement with an individual landholder, the process is less complicated. But when an association has to deal with ejido commissions representing the interests of a large group of landholders, the negotiations require both skill and a thorough knowledge of the situation. Such negotiations have always been undertaken by the elite *socios*.

In March 1948 Juan Barrientos, the president of El Carmen, sent a letter to the ejido commission in San Diego Chalma asking for permission to change the proposed route of the *galería* wells and to continue excavating in a straight line, which the well diggers believed would lead to the intersection of larger aquifers. The ejido commission in San Diego Chalma answered by letter that it could not possibly grant permission for the right-of-way without a financial settlement to pay for the "damage" to ejido lands as a result of the *galería* construction.

The *municipio* of Altepexi and surrounding fields (looking toward the south)

A typical irrigated cornfield in Altepexi

A typical street in Altepexi

A keeper of water and earth sorting her recently harvested tomatoes

A panoramic view of the southern Tehuacán Valley

A cornfield being irrigated, showing the elaborate infrastructure needed for sequential and uniform water distribution

A sluice gate (*compuerta*) for the regulation of water distribution

A main irrigation canal with a gate (*compuerta*) for a lateral going to a field

Mother and son: two keepers of water and earth

A new cement-lined canal

A campesino working to assure adequate irrigation for his field

A recently irrigated field

Regulating the flow of water: constant work and oversight

Preparation of a field for irrigation

Monitoring and controlling the flow of water

The Ladrón de la Huerta, guarded by members of Purísima No. 3

José Arribas, Juan Barrientos, Angel Santos, and Pedro Bartolomé met with the president of the ejido commission but were unable to reach an agreement. They felt the commission's request for eight pesos per meter of land crossed was unreasonable. According to the calculations made by José Arribas, El Carmen needed to cross at least five hundred meters, amounting to a payment of four thousand pesos for the excavation permit.

In an attempt to obtain government intervention on behalf of the association, Angel Santos proposed to name a three-member commission to go to Puebla, the state capital, and hire a lawyer. The lawyer would talk to officials in the Departamento de Asuntos Agrarios y Colonización who were in charge of the local ejido commissions. The object was to persuade the local commission in San Diego Chalma to decrease its outrageous demands. El Carmen members were in unanimous agreement with the plan. Another member further suggested that the commission be headed by José Arribas, because of his extensive experience in handling similar problems. Arribas, however, claimed that he had lost too much time in association activities. He countered by saying that there were many other members who should be obliged to participate in important association matters. Héctor González, a *socio mayor,* said that Arribas should not refuse to do what was asked and that the association had both faith and confidence in his abilities. Arribas was adamant in not going, so Ernesto Cofiño volunteered to go in his place. The membership reluctantly agreed. At Cofiño's request, they also agreed to pay a five peso per diem to compensate for time and money lost while away on association business.

On May 17, 1948, the association received a letter from the San Diego Chalma ejido commission granting permission for the *galería* construction; the special commission did not go to Puebla. Instead of a cash payment, El Carmen agreed to give twenty-four hours of water on the thirty-first day of the months with thirty-one days, and, in order to avoid interruption in the *tandas,* the San Diego Chalma ejido commission agreed to sell back to the association the twenty-four hours on January 31 and March 31 during non-leap years and on March 31 during leap years. The price for the water would be the going rate for the sale of equal *tandas.* The members accepted these terms and agreed to sign a formal agreement witnessed by the justices of the peace in both San Diego Chalma and Altepexi.

It was José Arribas who negotiated the terms of the agreement and eventually signed for the association. He had initiated the negotiations before refusing to go to Puebla. It is important to note that Arribas had both the power and the authority to negotiate for El Carmen without the knowledge of the members. It is doubtful that they would have named a special commission if they had known about his negotiations. Arribas's power to act on behalf of the association was never questioned by the other members, and the vote was automatically approved.

El Carmen's *galería* construction went smoothly for nearly a year, until the owner of La Huerta, Angel Herrera, initiated legal action against the association for beginning to excavate within the boundaries of his hacienda. Herrera charged that they had neither asked for nor received permission to excavate. In order to avoid the high cost and time involved in a legal action, the association decided to try for a negotiated settlement.

Pedro Bartolomé took charge and was able to reach a quick agreement. The terms, as requested by Herrera, were for the association to give him eight six-hour shares, amounting to a total of forty-eight hours of water per month. Herrera also agreed to pay the monthly charges as required of all members to cover association expenses.

In the three years that followed, Angel Herrera was slow to pay his dues. Frequently, the association had to send someone to ask him personally for the money, and his failure to pay often caused considerable delays in finishing *galería* cleaning and deepening projects. By January 1956, Herrera had obtained water for his land from other sources and sold his shares back to the association for a reportedly extravagant price of five to ten thousand pesos. Herrera had, in effect, been able to take El Carmen for a large sum, and some members claimed it was something he had planned from the beginning. Clearly, there was nothing the members could do about it, because the new wells were vital for the maintenance of the flow rate.

The *galería* system that is the source of water for P1 and P3 is one of the largest and longest in the Tehuacán Valley. The infiltration tunnels and breathing wells are under the city of Tehuacán, and they intersect with the surface at El Zotolín along the Tehuacán-Teotitlán highway just north of San Diego Chalma. From El Zotolín the water runs in an open canal, except for a short distance where an old and dry *galería* system is used as a subterranean canal, to the Ladrón de la Huerta, a distance of three kilometers. The Ladrón de la Huerta is a water-dividing mechanism, called a *medidor,* which separates the waters of P1 and P3.

The cleaning and maintenance of the *galerías* and the common canal are done jointly by the two associations. In 1963 the Peñafiel Bottling Company, having contracted to use the flow at El Zotolín for the generation of hydroelectric power, began sharing the cost of *galería* maintenance. According to the maintenance arrangement, Peñafiel would pay half of the total cost, the other half to be divided evenly between P1 and P3. The upkeep of the canals to the south of the Ladrón de la Huerta was the responsibility of the two irrigation associations.

The Ladrón de la Huerta is a large rectangular reservoir 3 meters wide by 6 meters long by 1.5 meters deep. The sides and bottom are made of concrete partially recessed into the ground, and the edges extend about 0.5 meters above the ground. The terrain on all sides is relatively flat. Water coming from El Zotolín runs into the northwest corner of the reservoir. On the opposite side is an opening that extends from the bottom of the reservoir to a height of 0.5 meters and that can be adjusted to widths of 0.5 and 1.0 meters. The size of this opening determines the flow of water going to the P3 canal system. On the east side of the reservoir is a spillway that extends along the entire 6-meter side and is approximately 0.3 meters below the top edge of the other three sides. This spillway provides the P 1 portion of the flow as excess runoff; the volume is determined by the variable size of the opening leading to the P3 canal.

The Ladrón de la Huerta was designed to divide the 300 lps total flow from the common canal into 200 lps for P3 and 100 lps for P1. If the total flow into the reservoir decreases, the flow to P1 trickles to a stop before the flow to P3 begins to be affected. The P1 flow, therefore, is an overflow, and any increase in the total flow goes to P1. This arrangement is part of the original 1946 sales agreement and is clearly stated in writing.

The common *galerías* require infrequent cleaning; only three major cleaning projects were undertaken between 1946 and 1973. On the other hand, the surface canals, in constant need of upkeep, are a continual financial drain on the associations. The cleaning and canal improvement activities of P3, as described below, are representative of how the associations maintain and improve their physical facilities.

Work on the surface canal system involves a number of distinct activities, each of which entails association planning, financing, and implementation: cleaning, a time-consuming chipping away of travertine buildup called *desensarre;* deepening of the canal floor (*repiso*); cement lining of canals (*revestimiento*); the renovation and rebuilding of old unused canals; and the construction of new canals.

The most frequent activity is canal cleaning. P3 is constantly cleaning one canal segment or another between El Zotolín, Las Seis, and Agua Ancha. Las Seis is where the main canal starts to fork into numerous smaller canals going to the individual fields; these small canals are maintained by the members who use them, and in most cases the costs are not covered by the association treasury. An exception is the canal from Las Seis to Agua Ancha, a distance of three kilometers; this canal is the main conduit of water separating the members who have land in El Alto, located above the town of Altepexi, from those below in the area known as La Mesa and Atzompa.

As mentioned above, the maintenance of the canal between El Zotolín and the Ladrón de la Huerta is a joint venture between P1 and P3, and both associations provide labor and share the cost equally. According to most members, a canal regularly conducting fast-flowing water should be cleaned at least once every two months. The criterion for cleaning is the rate of travertine buildup on the sides and bottom of the canal, which eventually leads to spillage and overflow. Furthermore, it is cheaper, easier, and faster to chip and scrape away relatively loose and porous travertine deposits before they thicken and harden. Members and governing officers often disagree on when canal cleaning should begin. Frequently, spillage and continual overflow prompt the beginning of a cleaning project, often as an emergency procedure. Canals are generally cleaned in one-kilometer sections, and a crew of eight usually finishes in ten working days, depending on the amount of travertine deposits.

In the period between 1945 and 1948 the canals were usually cleaned by the shareholders themselves. Beginning in 1948, P3 began to contract laborers to do the cleaning because of difficulties in getting the members to participate in labor details. By the end of 1956 the hired work crews had completely replaced the members. Association officers or specially appointed members continued to supervise the work.

The regular use of work crews also meant that the association had to ensure timely payments by all the members in order to meet the payroll. Laborers who were not paid on time usually did not return the next day. When the association had several ongoing projects, the income from the regular charges was often not sufficient to cover expenditures, and extra charges had to be imposed. Getting members to pay on time became an increasing problem. At times, work had to be stopped prematurely because the treasury (the *caja*) was empty. There are cases in which the elite *socios* stepped in to impose heavy fines and to make serious threats to sell *tandas* of members delinquent in payments. In most cases, the threats were quite effective, but constant tardiness often resulted in delays.

Deepening of the canal floor (*repiso*) is a longer and more expensive undertaking than cleaning and occurs at irregular intervals. A *repiso* is necessary when a canal floor has been raised by travertine deposits that cannot be removed by routine cleaning, although some *socios* claim that this can be prevented if the cleaning is thorough and frequent. In *repiso* the floor must be reexcavated using sledge hammers and chisels. The procedure usually means additional one-time charges of one to two pesos per hour of water owned by each *socio*.

In 1963, P3 made plans to construct a new canal parallel to one rented from P1, but in July 1963 it was denied permission for the project by the SRH and the *municipio* of Tehuacán. The SRH suggested to P3 that it consider getting permission to acquire an old and unused canal near the highway between Tehuacán and Altepexi. This canal had been used to conduct water to a sugar mill twelve kilometers south of Altepexi in the *municipio* of Calipan and belonged to a lawyer living in Mexico City. According to the SRH engineers, it would be a relatively simple matter for P3 to get permission to use the canal. The SRH engineers and the *municipio* officials felt that using the old canal would involve fewer negotiations for rights of way and would ultimately cause fewer problems.

After corresponding with the owner, the association received permission to use a section running from the entrance of La Huerta to the railroad crossing just north of Altepexi known as Viuda de Sánchez. Once permission was granted, however, a major renovation and restoration project had to be undertaken. Before the renovation and reconstruction could begin, the association also had to secure permission from the ejido commissions in both San Diego Chalma and San Pablo Tepetzingo, because the canal crossed ejido lands in both. Local landholders feared that the renewed use of an old canal could damage existing canals, cause water spills that would injure crops, and pose a constant nuisance due to debris from the renovation and subsequent cleaning. After an exchange of formal letters and a series of meetings between the association officers and the various officials, written permits were issued. The terms of the agreement called for cash payments of 1,500 pesos and 500 pesos to the respective ejido officials in San Pablo and San Diego. The renovation began in September 1963.

To pay for the new project and continue other routine maintenance, the association members were charged an additional 2 pesos per hour, collected at the end of August 1963. This amounted to 1,460 pesos, which, together with other association funds, was used to pay the lawyers who prepared the legal documents and the officials in San Pablo and San Diego. During the first two weeks of September, another 2 pesos per hour were needed. So far the total amount charged to the *socios* was 5,380 pesos, and two more 2-peso charges increased the total needed by 2,920 pesos to finish the project and pay for the inauguration celebrations held on December 8, 1963. The total cost of materials and labor for renovating 5.2 kilometers of canals was 7,200 pesos. The 1,100 peso difference between the actual cost and the amount paid by the *socios* was used, together with other funds, to pay ejido officials and cover other miscellaneous expenses.

Other maintenance projects undertaken by P3 include the installation and repair of gates that direct the flow to the proper *parcelas*. In 1945 most of the gates were wooden boards inserted into grooves in the sides of intersecting canals. These gates were gradually replaced by iron ones with hinges cemented into the sides of the canals. During the two decades between 1953 and 1973, P3 replaced about twenty wooden gates at a cost of 380 to 500 pesos each. The members felt that the investment was justified by the security provided by the iron gates. The gates are controlled by association personnel who open and close a padlock at the appropriate time. Although the gates are expensive and require constant cleaning in order to operate properly, they are very effective in preventing water theft and also assure that the members begin and end their *tandas* on time. Furthermore, the pressure of the water flow does not break the iron gates loose, sending water down the wrong canal.

The largest and most expensive maintenance and improvement project in P3 history was the *revestimiento* (cement lining of canals) from the Ladrón de la Huerta to Las Seis, a distance of approximately six kilometers. The P3 *socios* and association officers first began discussing the advantages of *revestimiento* at a meeting in December 1968. Angel Santos and Florencio Martínez pointed out that other irrigation associations in the Tehuacán Valley were saving on maintenance costs and losing much less water after *revestimiento* of the high-flow canals. Furthermore, as part of rural development in the Papaloapan Basin, which includes the Tehuacán Valley, the federal government had agreed to pay 50 percent of the total cost. The associations had to apply to the Papaloapan Commission in Tehuacán, the local branch of the SRH, for matching funds. The engineers from the commission had made the possibility of cost sharing known to all the irrigation associations in the valley.

At a P3 association meeting in January 1969, Angel Santos informed the members that he had met with the engineers, lawyers, and accountants from the Papaloapan Commission. The engineers had reported on their survey of the six-kilometer canal, and the accountants had made an estimate of 480,000 pesos. It was agreed that the government would pay a little more than 50 percent of the total cost. The accountants estimated that the cost to the association would be 148,000 pesos, or 31 percent of the total expenses. Completion time was estimated to be approximately six months provided the capital, labor, and materials were available. It was, therefore, up to the association to decide whether and when it wanted to begin the *revestimiento* project. Over two years passed before the project got under way. It began in August 1971 and took nine months to complete, costing the association approximately 190,000 pesos.

According to the original cost estimate of 148,000 pesos, the project would have cost the *socios* 202 pesos per hour of water owned, with the payments spread over a six-month period. This amounted to a little less than 8 pesos per hour per week or, because P3 has three-hour hectare-shares, 24 pesos a week per ha-s. In reality the total was a little lower, because part of the cost was offset by income from fines, water sales, rent from hydroelectric plants, and other miscellaneous sources. According to the financial records, the cost to the membership based on the number of hours of water owned was 125,560 pesos.

The payments were made by the members in the following manner: beginning on August 1, 1971, the members were charged three pesos per hour per week. This lasted until the end of the second week of October, when the weekly payments were raised to five pesos per hour. These payments were made until the end of April 1972. The project was officially completed on May 1, 1972, when the membership resumed making regular two-pesos-per-hour monthly payments.

Upon completion of the *revestimiento,* the association had paid 55 percent of the total construction cost, amounting to 190,000 pesos, and the government had paid 158,000 pesos, or only 45 percent. The total cost of approximately 348,000 pesos was less than estimated, but the government had contributed less than 50 percent and P3 more. Since the completion of the *revestimiento,* there have been suggestions and discussions of cement-lining other main canals, but by the end of 1983, no new projects of this type had been planned.

Cooperation between P1 and P3 to maintain the *galerías* under Tehuacán has a long and difficult history. The following is a reconstruction of some of the major events and conflicts. In July 1948 the president of P1 sent a letter to P3 asking for help and cooperation in cleaning the *galerías* under Tehuacán. According to the elite and more knowledgeable P3 members, this was the first *galería* cleaning project since the purchase of the system. Instead of agreeing to cooperate, the governing officers and members of P3 decided that the canal between El Zotolín and the Ladrón de la Huerta was in more urgent need of cleaning. P3 did name a special commission to negotiate an agreement with P1 to begin the *galería* project at a future date.

At the end of January 1949, P1 sent another letter urgently asking that they begin the *galería* cleaning project. The P3 members finally agreed to make the financial commitment to begin the project, voting to charge only 1 peso per hour, or 730 pesos, initially. In April they voted an additional peso per hour to continue the cleaning.

The slow and difficult job of *galería* cleaning continued into the summer of 1950. It was done by a four-man crew working much like the *galería* construction crews described in Chapter 4. In August another peso-per-hour charge was made to continue the work. By November the members were told that the work would have to stop because of lack of funds. The association officers decided they could raise additional funds by collecting outstanding fines from those who were not helping in the project and from those who did not bother coming to association meetings. This solution was unanimously approved. The outstanding fines amounted to about the same as an extra one-peso charge. It was decided to make the debtors pay up before new charges were made on the rest of the members.

By January 1951 the largest section of the *galería* had been cleaned and deepened, and the associations agreed to clean the other, shorter, arm of the *galería* network. The entire project was completed by the middle of July 1951; it had taken almost two years. Because of the incomplete records kept by P3, it has been difficult to calculate the actual cost of the entire cleaning operation. The financial records indicate that 1-peso-per-hour contributions were made on five separate occasions, amounting to 3,650 pesos over two years. The total cost would then have been 7,300 pesos, since P1 paid 50 percent. In actuality, the cost must have been somewhat higher, because a few of the members contributed labor instead of money.

The next *galería* cleaning project was done more than ten years later and began in December 1962. This project was a joint venture between P1, P3, and the Peñafiel Bottling Company. Peñafiel agreed to pay 50 percent of the cleaning cost, because the constant bickering between P1 and P3 caused serious delays in the implementation of long-needed *galería* cleaning. Officials at Peñafiel felt that an investment of five thousand pesos was justified to ensure the continuous and efficient operation of the San Andrés hydroelectric plant. With the leadership and persuasion of the Peñafiel engineers, the project was completed in six months, and the water flow was restored to 300 lps.

The next *galería* cleaning was in the latter part of 1971. This time Peñafiel took charge of the entire operation; it contracted and paid for all the labor needed to complete the project. Company officials said they had neither the time nor the interest to wait for the associations to make up their minds on when to begin. Peñafiel did not disclose the cost of the cleaning to the associations, and the members did not make any financial contributions.

In order to maintain the 300 lps flow rate at El Zotolín, the *galerías* were cleaned three times in the period between 1945 and 1973. During the ten years between cleaning projects there was a gradual decline in the flow rate for P1. Furthermore, there were significant short-run fluctuations in the flow rate, making it difficult to document a gradual decrease on a daily, weekly, or even monthly basis. The fluctuations were caused by a pulsating and uneven flow from the aquifer, and accurate flow measurements (*aforos*) were very difficult to make.

The measurement of the flow rates had been done sporadically by P1, P3, and the Papaloapan Commission. The most reliable measurements were made by qualified engineers who obtained the only results that were acceptable to everyone concerned. The P1 members were always the first to make note of the decreases, and they had to decide whether a decrease was due to a short-run fluctuation or a long-term trend.

The frequency of the joint maintenance projects by P1 and P3 was tied to the changes in the *galería* flow rate and the condition of the canal between El Zotolín and the Ladrón de la Huerta. The fact that the P3 flow was not affected by decreases in the flow rate has been blamed for P3's lack of interest in frequent maintenance. The members themselves, however, had neither the training nor the equipment to measure the flow rates. Their estimates were based on the depth of the water in the canals, which only gave a rough idea of relative differences in the flow. Consequently, both associations were constantly calling for *aforos* by SRH engineers, who, at times, were reluctant to come.

CONFLICT BETWEEN P1 AND P3

Fluctuation in water flow and the mechanics of division at the Ladrón de la Huerta have been the principal causes of conflict between P1 and P3. Virtually all interaction between the two associations has been related to water division. P1 had repeatedly attempted to change the overflow arrangement to a proportional division that would result in equal decreases and increases in the flow to both associations when there were changes in the total flow. Events between 1946 and 1973 illustrate P1's efforts to make a change and include official correspondence with P3, meetings between the two associations, and intervention by third parties.

The original *galería* sales agreement stipulated the joint ownership, administration, maintenance, and division of water as previously described. No water could be taken by either association from any place on the common canal between the Ladrón and El Zotolín. During the summer of 1948, some P3 members began to suspect that certain P1 members were illegally tapping the canal above the Ladrón. The P3 leaders contacted the administrators of the Tehuacán Electric Company and asked them to investigate the possible infractions. The appeal to the Electric Company was made because of its longstanding agreement with both associations to use the total flow to drive the turbines of the San Andrés plant, located just below El Zotolín. As part of the agreement, the company was responsible for maintaining the canal system near El Zotolín as well as guarding the canals to prevent loss and theft in the areas where P1 was suspected of illegally taking water.

In November 1948 the president of P3 sent a letter to the Tehuacán Electric Company formally accusing P1 of taking water from above the Ladrón. The letter stated that P1 was taking water with impunity and asked the Electric Company to intervene. A copy of the letter was sent to P1. A month later all three parties sat down and negotiated, with the Electric Company leading the discussions. The outcome was that only the P1 members who had land above the Ladrón would be permitted, under the supervision of the Electric Company's *aguador,* to take water there during their *tandas.* Although P3 would have preferred that absolutely no one take water above the Ladrón, they reluctantly agreed to permit three P1 shareholders to do so. The P3 governing officers still believed that P1 was taking more water than it was entitled to.

The accusations and conflict between the two associations continued during the subsequent *galería* cleaning project, which was completed in 1951. Afterwards, there were no records of any formal conflicts until November 1956. At this time, P1 initiated efforts to change the system of water division at the Ladrón de la Huerta.

On November 20, 1956, a letter from the Papaloapan Commission's office in Tehuacán was read to the P3 general assembly. The letter from the chief engineer and administrator instructed association leaders or their representatives to attend a meeting in Tehuacán on November 22, and to bring the necessary documents to demonstrate the basis for the water division at the Ladrón. The president of P3 proposed that the association name a commission to go to the meeting in Tehuacán, to find out why the letter had been sent to P3, and to determine who had been responsible for provoking the intervention of the Papaloapan Commission.

Three weeks later, the president of P3 informed the members that P1 had complained to the Papaloapan Commission that P3 was taking too much water at the Ladrón. The P1 members felt that the aperture from the Ladrón to the P3 canal was too large for one of two reasons: part of the metal plate had corroded away because of the rapid water flow, or the dimensions had been changed willfully. The P1 members were convinced of the latter; P3 claimed it had not touched anything. P1 asked the SRH to intervene by making accurate *aforos* for the flow going to both associations. The P3 members perceived an immediate threat to the overflow arrangement and also felt that they were being unreasonably attacked because the flow to P3 never increased when the total flow did; the increases went to P1. As a result, P3 felt compelled to organize yet another special commission, to protect its interests and to represent the association before the SRH.

A month later it was discovered that a P3 shareholder was responsible for the current problems with P1. This member had shares in both associations (his was the only case of overlapping membership), and he had persuaded P1 to go to the Papaloapan Commission. Since the majority of his shares were in P1, he would have benefited from the government's giving more water to P1. Understandably, the other P3 members became very hostile toward him, and a year later he sold his shares in P3.

On May 16, 1957, at the SRH offices in Tehuacán, a meeting was held between the special commission from P3, the president and secretary from P1, and the chief engineer at the Papaloapan Commission, who represented the SRH. The representatives from P1 claimed that their water had decreased considerably because the Ladrón de la Huerta was not dividing the water properly. In effect P1 was asking the government engineer to change the written agreement for the division of water at the Ladrón.

Angel Santos, the president of the special commission from P3, proposed that the Ladrón be completely reconstructed according to the original specifications in the purchase agreement. He said that the water that went to P1 was, after all, an overflow and should not be a fixed measurement. A heated discussion followed. At the end, the proposal was approved and all the parties concerned agreed to meet again with the engineers after the *aforos* had been made.

On June 6 and 11, the engineers from the Papaloapan Commission made two *aforos* at the Ladrón. The results were inconclusive because of the constant fluctuation in the flow. Consequently, the engineers decided to continue taking measurements during an entire month in order to get an average measurement of the flow in both canals. The next measurement was scheduled for June 19.

The P3 members were sure that they could not lose, because it was common knowledge and perfectly legal that the P1 water was the overflow and that P3 had been getting the same amount since the Ladrón had been repaired many years earlier. The entire P3 membership unanimously agreed that every member should be present at the Ladrón on June 19 to show support and witness the *aforo*. It also decided to impose a fifteen-peso fine on any member who did not attend.

The *aforos* made on June 19 and in successive weeks during July were very erratic because of variation in the flow. According to the SRH engineers, the flow to P3 varied from 220 to 185 lps, and to P1 from 105 to 80 lps. These figures were neither declared by the engineers to be official nor recorded at the Papaloapan Commission offices, thus casting serious doubt on their accuracy.

The final outcome of several months of periodic *aforos* and numerous meetings between the associations and the government engineers was a decision to reconstruct the Ladrón so that the split would be 200 lps and 100 lps for P3 and P1, respectively. Reconstruction did not mean that they would demolish the entire structure, but they would have to rebuild the spillway for the P1 overflow and change the P3 outlet dimensions according to the revised specifications. Because the total flow was often 25 lps below the normal 300 lps flow rate, the chief engineer also strongly recommended that the associations promptly begin cleaning the *galerías* in order to restore the flow rate.

Nothing immediate was done to reconstruct the Ladrón or to clean the *galerías*. Instead of the expensive reconstruction, a slightly larger metal plate was put into one side of the P3 outlet, reducing its flow and adding about 20 lps to the P1 flow. This amounted to an implicit admission by P3 that its flow rate had been excessive, and temporarily alleviated the conflict.

When P1 asked for more *aforos*, over two years had passed and the flow rate had once again decreased. The SRH asked the governing officers of P3 to be present at a meeting in Tehuacán on December 22, 1959. In the meantime, the P3 membership approved a proposal to begin the *galería* cleaning and the reconstruction of the Ladrón de la Huerta. The representatives of P1 and P3 met with the engineers from SRH, but nothing substantial was settled. The two associations continued to argue over the division of water at the Ladrón. The periodic *aforos* exacerbated the situation, because the short-run variations in flow made them unreliable, hence useless for resolving the dispute.

By the end of 1961 neither the construction plans for the Ladrón nor the actual physical measurements had been made. The plans were finally completed and work began on January 18, 1962. No accurate figures are available on the cost of the project or on how the expenses were divided between P1, P3, and the SRH. The reconstruction was followed by the cleaning of the *galerías*, which began in the summer of 1963.

During this time, P1 informed P3 that the latter could no longer continue to use the canal owned by P1 to conduct water from the Ladrón to Las Seis. P3 had used the P1 canal for about eighteen years, but, after some delay, it acquired and renovated an old, unused canal, as described previously.

On June 6, 1963, the governing officers of P3 were notified to attend a meeting the following week at the Ministerio Público in Tehuacán. There the P3 representatives were informed that P1 was formally accusing P3 of continuing to take too much water, despite the reconstruction of the Ladrón. The ministry lawyer said that the associations should obtain a reliable measure of the amount of water that P3 was taking and be compelled to make a new legal agreement on the water division. The representatives from both associations reluctantly agreed to take flow measurements at the Ladrón with all the members from both associations present as witnesses.

On June 17, 1963, a lawyer representing P3 in its dispute with P1 came to Altepexi and informed the association that, according to the calculations made by the engineers, P3 was taking too much water. This was stated clearly and emphatically in the final report to the Ministerio Público. As a result, the ministry ordered the new adjustable opening of the P3 gate in the Ladrón to be reduced to 30 by 32.5 centimeters in order to give a flow of 200 lps. The overflow would continue to go to P1 as stipulated in the original purchase agreement. The lawyer also informed the association that its cost for the legal work and the private engineers who did the *aforos* was six thousand pesos.

At the next P3 association meeting, the president told the members about the conclusions reached by the engineers and the Ministerio Público. When members were told about the high costs, a lengthy and heated discussion followed. Eventually the members agreed to pay eight pesos an hour, to be paid over four weeks at two pesos per hour. They also named a special commission to be present at the formal signing of the agreement between P1 and P3. José Arribas was named the commission leader.

The dispute over the Ladrón had involved two government bureaucracies, the SRH and the Ministerio Público. The engineers from the SRH could only make measurements and try to enforce the terms of a preexisting agreement, the 1946 sales contract. Once the *aforos* had been made and the two associations had agreed to the terms, any disagreement over compliance was in effect an infraction subject to legal sanction. P1 had accused P3 of not complying with the terms of the agreement and asked for legal intervention by the ministry. The additional *aforos* were part of a new investigation by the ministry lawyers and the association representatives. P3 was subsequently found guilty of noncompliance, which resulted in the revision of the P3 flow.

The following incident is a further example of the inability of P1 and P3 to agree even on minor matters. On November 10, 1963, P3 received a carbon copy of a letter sent to P1 by the Peñafiel Bottling Company. In the letter, Peñafiel discussed the need for and fairness of a 50 percent pay raise for the workers currently cleaning the *galerías*. P1 agreed to give the workers a raise but stipulated that the cost of the raise be shared proportionally between P1 and P3, according to the proportional division of the water flow.

The governing officers of P3 felt that if P1 and Peñafiel agreed to the raise, then P3 should also agree. But they also felt that the new water increase that P1 was receiving should be taken into account, because the P3 flow was unchanged. The current arrangement was for P3 to pay 30 percent, P1, 20 percent, and Peñafiel, 50 percent of the expenses. The president of P3 suggested that the percentages paid by P3 and P1 be reversed, or that P1 divide the increased water volume evenly with P3.

In a letter dated December 16, 1963, P1 refused to start paying 30 percent of the workers' raise, because it claimed there had been no increase in the water flow. Even if there had been an increase, it would only partially make up for the volume that P1 had not gotten over the years. P1 went on to say that P3 should pay for the entire increase in the workers' wages. Wishing to avoid further problems with P1 over the division of the water flow, P3 agreed.

During this time, the SRH chief engineer became aware of the problems between the two associations. He sent letters to both associations asking that they demonstrate some good will and that, for once, they try to cooperate. After having been found guilty of noncompliance, the P3 leaders chose not to provoke more problems. They felt that a small increase in the workers' pay was a good investment toward continued tranquillity. The warning from SRH was timely and prompted P3 to pay and not stir up further conflict.

From 1964 until the middle of 1971, there was only minor bickering between the two associations. During this time there was a relatively constant and high flow in the *galerías* primarily because of Peñafiel's increased *galería* maintenance efforts. On several occasions Peñafiel had hired workers to clean specific sections of the *galería* where the infiltration was decreasing. On these occasions Peñafiel did the work and covered the expenses without informing either of the associations. Its principal motivation was to keep a good flow of water in its hydroelectric plant, which also resulted in a higher flow for the two associations.

In August 1971 the P3 shareholders began noticing a flow of water that was considerably greater than any that they could remember. Peñafiel had just completed a major *galería* cleaning project, which was probably the reason for the increased flow. The P3 governing officers hired an engineer to measure the flow, and the results were a total flow of 380 lps: 277 lps to P3 and 96 lps to P1 (the discrepancy between the total and the partial flows was due to errors in the measurements). Florencio Martínez claimed that there must have been an error somewhere, because after the *galería* cleaning in 1963, P3 had 217 lps. José Arribas suggested that they notify the authorities. The membership decided to take no action, but instead to name a special commission and wait for P1 to send a letter of accusation.

The letter never came, but on March 25, 1972, a visitor arrived from the executive branch of the federal government (Dirección General de Gobernación de la Presidencia Municipal). At P1's request, he came to investigate the measurement and division of water between the two associations, since P1 claimed to be missing a substantial portion of the increased flow to which it was entitled. The P3 governing officers retorted that, since the *galerías* were purchased, there had been an equitable division of water, proven repeatedly over the past twenty years. No mention was made of the 1963 Ministerio Público intervention. At the next meeting of the P3 members, they decided to contact a lawyer, who would be able to determine whether the verbal message from Gobernación had any legal standing. The assembly went on to select a special commission to represent the association in the new conflict with P1. The commission members included Arribas, Florencio Martínez, Santos, and three other *socios mayores*.

On May 18, 1972, the president of P3 received a telegram from the private secretary to the governor of the state of Puebla. The telegram informed him of a meeting at the governor's office in Puebla on May 23, which the special commission was required to attend. Once again, P1 had accused P3 of taking too much water. It had requested the governor's intervention, in addition to new flow measurements and a permanent proportional division of the water. This marked the first time that P1 had appealed to the state governor to alter the terms of the *galería* purchase agreement.

At the meeting in Puebla, Angel Santos said that, although P3 did receive more water, the members from P1 knew that this was the legal arrangement as verified by SRH and the Ministerio Público in Tehuacán. After an extended discussion, the governor's secretary asked P1 to put the petition in writing so that the governor could resolve the problem. As of the middle of 1981 nothing had been decided. P3 continued to receive a fairly reliable flow of at least 200 lps, sometimes considerably more, and P1 continued to be affected by the short- and long-run variations in *galería* flow.

The conflict between the two associations began in 1948 and has continued to the present. The Ladrón still provides an overflow to P1, and the controversies have continually centered around the size of the opening to the P3 canal. It was apparent from the beginning that the conflict could not be resolved by the associations themselves.

The conflict has involved three government ministries, and despite lengthy, often expensive efforts, no solution has been reached. The first ministry to be involved was the SRH, by providing technical expertise in flow measurements to determine whether the terms of the purchase agreement were being followed. Then the Ministerio Público, representing the Ministry of Justice, enforced the terms of the agreement. The third intervention was from the executive branch, the Secretaría de la Presidencia, acting through the governor of Puebla. The governor, with subsequent presidential approval, had the power to authorize contractual changes, but in the case of the petition by P1, no action had been taken at the time of this writing.

This case demonstrates the government's reluctance to alter the terms of private contractual agreements, even when those terms affect the use and distribution of water, which is the property of the state.

PURÍSIMA NO. 3 AND THE TEHUACÁN ELECTRIC COMPANY

The negotiations between the Tehuacán Electric Company and P3, described below, show the relative positions and influence of the two groups. The negotiations illustrate how a group of small agriculturalists went to great lengths and expense in attempting to change a longstanding contractual agreement for the use of *galería* water.

The Tehuacán Electric Company generated electricity for the city of Tehuacán and the surrounding communities in the period between 1927 and 1952. Its hydroelectric generators were driven by the water from the *galerías* of La Purísima Nos. 1 and 3 and several smaller *galería* systems, as well as the natural spring of San Lorenzo. The flow rates from each source were measured, joined, used to run three hydroelectric generators, and then separated into the original component parts; real and alleged water loss during the process of use frequently caused conflict. The Electric Company paid a user fee to the owners of the water systems according to agreements made in 1926.

Soon after the agriculturalists in Altepexi had purchased P3, the members tried to increase the user fee. In October 1946, P3 sent a letter to the Electric Company asking for an increase in the monthly payments from five hundred pesos to eight hundred pesos and an increase from twenty pesos to forty pesos for the canal cleaning. The Electric Company did not acknowledge receiving the letter. Nothing further happened until the association officers contacted legal counsel to meet with the administration of the Electric Company.

In June 1947 Javier Martínez-Fraile, the association's attorney, told P3 that, since the Electric Company had sold all three hydroelectric plants, there was no reason to pursue the matter any further. He did not say when the sale would go into effect. Pedro Bartolomé felt that P3 should get another lawyer, because it was obvious that Martínez-Fraile was not to be trusted. The association officers then asked Carlos Saavedra, an attorney who was related to two of the members, to talk to the director of the Electric Company. Saavedra requested the original contract between the association and the company, but the document no longer existed.

The lawyer suggested that P3 find an engineer to make accurate *aforos* of the water flow in order to confirm, without a doubt, the alleged water losses from the canals near the hydroelectric plants. He also recommended that P3 become a civil association (*sociedad civil*),[1] which he felt was a necessary prerequisite for any appeal and subsequent negotiation with the administration of the Electric Company and other civil authorities. A legally constituted organization, rather than a group of agriculturalists, would be in a better position to make claims if there had been a demonstrable loss. It would also be in a better position to demand an increase in payments. The association representatives, Miguel García, Pedro Jordán, and Pedro Bartolomé, agreed that the lawyer should proceed with the paperwork to establish a *sociedad civil*.

At a meeting on September 8, 1947, Miguel García informed the members that the lawyer had completed the legal papers. Rafael Salazar, the P3 treasurer, said the estimated lawyer's fee was 276 pesos, which should either come from the treasury or from an additional charge to each member. Although the lawyer had suggested that each member pay 3 pesos, Pedro Bartolomé felt that most of the members had only three to nine hours of water and should therefore pay in proportion to the hours owned. After much discussion, they decided that there was enough money in the treasury to pay the fee. On October 20, 1947, the paperwork was completed, and all that remained to make the *sociedad civil* official was the signatures of the association officers.

At about the same time the association received a letter from Rafael Prieto, a P3 shareholder, informing them that, according to a reliable source, the Electric Company would no longer need association water after January 1948. The letter went on to say that it was therefore unnecessary to continue negotiations with the company, and that it would be more important to repair the canals bypassing the hydroelectric plant. Furthermore, it was no longer necessary for the association to continue with the expensive paperwork to become a *sociedad civil.* Angel Santos stated that, even though the water would not pass through the hydroelectric plant, the legal papers might be useful in the future. Florencio Meléndez added that the association leaders should sign the papers because they had already incurred the lawyer's fees. The members unanimously agreed.

At an association meeting in December 1947, Martín Iglesias proposed that they meet with the Electric Company to determine exactly when it would stop using the water. If the company continued to use the water, the association should still insist on a new contract. Members agreed, and on January 14, 1948, officers met with the administrator of the Electric Company. He told them that the company would continue to use the water and that he would inform them by letter when it would no longer be needed. Regarding the requested fee increase, he said that the association should put it in writing so that he could forward it to his superiors in Mexico City.

At the next meeting of the *socios* on January 19, 1948, Clemente Alvarado insisted that they try to get compensation from the company for the continued water loss. Angel Santos said that, before the association could insist on anything, it must have reliable measurements as to the exact amount of water entering the hydroelectric plant and the amount leaving. The membership agreed with Santos and decided to employ an engineer to make the measurements.

In the middle of October 1950, the Electric Company verbally informed the association that the water flow through the Carta Blanca and Humilladero hydroelectric plants was no longer needed, because the plants were finally being shut down. Only the San Andrés plant would continue to operate, but it had been sold to a textile mill called San Juan A. Xaltepec, located in the *municipio* of Altepexi. They should collect rent for the water from Carlos González, the administrator of the mill.

At the association meeting later in October, Rafael Prieto proposed that the association negotiate a new contract with the mill, stipulating that it pay the new rates they had been trying to charge the Electric Company. Miguel García emphasized that the new contract should guarantee that no water would be lost in the canals or in the turbines of the hydroelectric plant; Gregorio Ortega suggested that the textile mill be obliged to adjust the water dividers if there were any decrease in the amount going to P3; and Martín Iglesias proposed that they construct a new water divider that would assure an accurate division of water for the associations. Florencio Martínez, together with Angel Santos, strongly supported this proposition, which was immediately approved by the membership.

The new water divider was constructed at the beginning of 1951, and the association was able to negotiate and sign a new contract with the textile mill that included a rate increase of 840 pesos per month.

The textile mill used the water without incident until March 1955, when labor unrest led to a strike by textile workers. At this time, Industrias de Puebla, the owners of the mill, informed the association that, because of the strike and the uncertainty of when it would be settled, they were going to close the hydroelectric plant. They told each of the four irrigation associations to use canals that bypassed the hydroelectric plant; they would inform the associations when they needed water again. The P3 *socios* decided to renovate a canal near the old road to Oaxaca.

The San Andrés hydroelectric plant did not, however, go out of operation. The Peñafiel Bottling Company had been using part of the electricity from the plant since 1927, and it decided to continue operating the plant, because it was a cheaper source of electricity than the Federal Electric Commission (Comisión Federal de Electricidad).

Peñafiel agreed to pay the same monthly users' fee as the mill and eventually became actively involved in *galería* maintenance, because a constant and reliable *galería* flow was required for generating an adequate number of kilowatts for the bottling plant. Neither P1 nor P3 requested any changes in the monthly payments, since the members were well aware of the expenses Peñafiel had incurred for *galería* maintenance.

In late 1973 the hydroelectric turbines were damaged by a severe earthquake. Peñafiel decided not to repair the facilities, and hydroelectric generation with *galería* water came to an abrupt end. The *galería* systems had not been damaged, and P1 and P3 were once again responsible for all the maintenance work.

The relations between P3, the Electric Company, the mill, and Peñafiel, show that a *sociedad civil* of agriculturalists is, except for a few special cases, relatively powerless in trying to increase the charges for water rental. The associations probably had no choice about their water being used in the hydroelectric plants, because no one ever suggested diverting the flow to bypass the generators and holding out for more money. The use of the water in the plants predated the *galería* associations in Altepexi, and the former owners of P3 had made the agreements with the Electric Company, which the P3 members were unable to change.

SUMMARY

The activities and events described in this chapter show that particular *socios* have been instrumental in making decisions about leadership selection, expenditures, the timing of maintenance and construction, and relations with outside individuals and organizations. The most visible and vocal *socios* have been Angel Santos, José Arribas, and, to a lesser extent, Florencio Martínez. These elite *socios* appear to have the complete support of the rest of the *socios* and at no time has there been any evidence of a challenge to their leadership. Therefore, any reference to particular association plans, decisions, policies, or actions is, in effect, a reference to these elite members. Very little has occurred without their direct involvement and knowledge.

All association officers and special commission members have been *socios* from the *municipio,* and there is no evidence of outside direction; the internal leadership structure and selection process is strictly a local phenomenon. The elite members have been able to change the criteria for filling the routine positions and have determined when they would intervene as the leaders of special crisis commissions.

The Decree of April 1954, which prohibited the sale or rental of water to nonmembers, demonstrated the association leadership's power to control the disposition of each member's shares. Although provisions of the decree, enforced by the threat of a large fine, were imposed only once, it was an effective means of restricting water use in response to external threats and agitation.

The leadership has authority over the admission of new members by its absolute control over the *tanda* sequence; that is, leaders have the power to decide when a new member will begin to receive water. Absolute control over the *tandas* is also an effective means of enforcing timely payments by the members, since those with longstanding delinquent payments are faced with the sale of their *tandas*.

There is a common belief that the irrigation associations are inherently wealthy and can therefore be asked for money, as was illustrated by requests to P3 from the *municipio* of Altepexi, the church, and an individual shareholder. The expense of cleaning and construction projects, however, means that the associations are constantly short of cash and must usually raise money by charging the members an appropriate amount per hour of water owned; the same procedure is followed when the association makes a contribution. The decision to give appears to depend on current expenditures and ongoing projects.

The formal procedure for dismissing the *aguador* not only demonstrates the seriousness of having an unreliable and felonious employee but also shows that the Mexican labor laws have to be followed. It was not enough to show cause for dismissal; the allegations had to be proven in order to comply with the requirements of the statutes cited.

The construction and maintenance projects completed by P3 and El Carmen demonstrate that the associations are able to collect large amounts of capital. The implementation of the projects shows that the association leaders have had to get permission from SRH to enter the *galerías*, negotiate with ejido commissions for the right-of-way, settle with ranch owners for digging rights in exchange for water, get permission from owners of unused canals, and seek cooperation from P1 for cleaning of common canals and *galerías*. The SRH has ultimate control over what the associations do, but few restrictions have been placed on the association's use of groundwater. An alleged prohibition on new *galería* construction has never been enforced.

The conflict between P1 and P3 shows the relative domains of the SRH, the Ministerio Público, and the office of the governor of Puebla. The SRH regulated access only to groundwater, and the *aforos* were used to confirm or deny compliance with the terms of the purchase agreement. The SRH did not attempt to change the terms of the agreement. The Ministerio Público tried to mediate the dispute, but did not alter the terms of the agreement either. The governor's office, with the power to allocate water to ejido land users, took the matter under consideration but did not make a decision.

The P3 irrigation association, whose operational history has been described above, has a differentiated internal leadership or elite. Their influence beyond the local irrigation association is limited, as demonstrated by the inability to reach agreements with other associations and, in particular, their inability to secure an increase in rates charged the Electric Company. The Electric Company and its personnel were connected, in a variety of ways, to higher levels of the government institutional structure and must have considered the association efforts as not much more than a nuisance.

The most important conclusion to be reached from the data presented in this chapter is that the internal affairs of the irrigation association are controlled by a small number of elite members. Although these members have at times held formal positions, their influence has been continuous regardless of officeholding. Their most important power lies in their ability to exercise control over the *tandas*. This control has been the underlying basis for compliance in raising funds for association operation and projects. The central function of the association has been to concentrate capital for expenditures and projects to maintain and increase the security of a constant water flow. All internal and external activities are related to this primary function.

7.

"WE ARE ALL CAMPESINOS": THE CONTRADICTIONS OF GROWTH

THE SHIFT from subsistence farming to cash crops in the Tehuacán Valley in the 1930s created competition for already-scarce irrigation water. As the price of water rose rapidly, many peasants who had shifted to cash crops were forced to borrow money to pay for water. Some incurred debts they could not pay and lost the land they had used as collateral. Commercialization enhanced the power of families that had been powerful before the agrarian reform and had managed to buy water from the neighboring haciendas that were divided during the reform. At that time, the landless workers were mobilized but their union was subsequently broken.

This chapter examines the structure of domination and the factors that undercut the ability of campesinos to mobilize effectively in order to gain access to land, water, and other desired inputs or services. These factors are critical not only for the Tehuacán Valley but for Mexico in general. Apparent acquiescence in the status quo or political conservatism has been attributed by some scholars to largely psychological reasons. In contrast, our analysis of the Tehuacán Valley combines ideological and structural factors.

The 1930s was a period of great rural ferment in Tehuacán as well as in many other areas of Mexico. In San Gabriel Chilac, Ajalpan, and Altepexi, unions were formed by landless workers who labored on the sugar plantations. These village-based unions had multiple purposes, including increasing wages on the plantations and improving the communities through cooperative stores and schools. In Chilac the union formed night schools, a cooperative store, and a political party. The unions and several other peasant groups that were petitioning for land were opposed within their own communities by coalitions of the local petty bourgeoisie who owned land and water. These individuals viewed the union as a threat to their own power as well as to their desire to acquire new land and water.

During the 1930s the union in Chilac pressured for land reform. It was finally successful when the government confiscated the land of Hacienda San Andrés, totaling more than 100,000 hectares that stretched across the valley north of Chilac. Although union members were allocated land from the hacienda to form an ejido, the hacienda owner thwarted the union's efforts to gain control of the water previously used to irrigate the hacienda lands.

Without water for irrigation, the land could not be cultivated. The hacienda owner managed to bribe the government surveyors into declaring the newly designated ejido holdings as nonirrigated land, meaning that the water rights did not have to be transferred with the land. The union members were stunned by the decision, realizing that their futures hinged on gaining access to water. After a series of unsuccessful petitions to the governor for water, the *ejidatarios* offered to sharecrop the land if the hacienda owner would provide the water (Carta de la Agrupación Agraria de San Gabriel Chilac 21 de julio 1932).

The hacienda owner, bitter over losing his land and worried that the relationship would legitimize the union members' need for and subsequent right to water, turned down the request. This provoked an angry reaction from the *ejidatarios,* who threatened to take the water by force. Realizing that his situation was tenuous and that the government could still confiscate his water rights, the hacienda owner decided to sell his water. The highest bid was made by an association of people from Chilac and Altepexi; the members were not the new *ejidatarios* but wealthy agriculturalists.

In 1948 the Sociedad Chilac-Tlacoxolco-Tititlán, which had purchased the water rights from Hacienda San Andrés, bought the remaining 192 hours of water controlled by the owner of the hacienda. The ejido members desperately tried to raise money to buy the water rights, but they were unsuccessful. As one of the *ejidatarios* explains, "We knew it was our last chance to get water but we were only poor peasants. None of us had money. We asked the government for help or at least a loan, but it took too long. Those who had money bought the water so they could make money. All we wanted was enough water to plant the land to feed our families. Those of us who shared this experience are still bitter."

The conflict over water was not solely between groups within the same community but between communities as well. This pattern of conflict pitting community against community contributed to the political control of regional caciques. As documented by the history of water conflicts, government vacillation on who had legal rights to the water contributed to suspicion and conflict between communities.

In 1931 the ejido in Chilac received an additional 394 hectares of land when the government confiscated Hacienda Venta Negra. Land from the hacienda was also given to four other ejidos in the region. In 1933 the Chilac ejido petitioned for water from the spring Atzompa, previously used to irrigate hacienda lands. The governor of Puebla had provisionally awarded the water to the community of San José Miahuatlán, which had been buying water from Hacienda Venta Negra before the agrarian reform. However, since there was no ejido in San José, the governor reneged and gave the water to the ejido in Altepexi. In 1936, by presidential decree, the ejido in Chilac was awarded the water from Atzompa. But in December 1936 an armed band of men from San José took over the irrigation canal and the dividers that sent water to the respective villages. The Chileños complained to authorities in Tehuacán, who sent troops that captured six men while the rest escaped. In the ensuing court case, San José made a new request for water.

In 1945 a water society was formed in San José, which bought rights to twelve days of Atzompa water from Hacienda Venta Negra. The ejido of Chilac kept eleven days and sixteen hours of water a month from Atzompa, but the quantity of water (36.7 lps) was only sufficient to cultivate 48 hectares of land divided between 144 *ejidatarios*. In 1954 a new struggle between the owners of water evolved as the spring produced less and less water and eventually went dry.

The conflict over water within and between communities of the valley highlights the divisions in the population. The genesis of many of the water conflicts in the Tehuacán Valley can be traced to decisions made by government officials living outside of the region. Equally important, by allowing private parties to purchase water rights after the agrarian reform, the federal government destroyed the political and economic viability of some of the ejidos directly affected. The result was to strengthen greatly the position of an already significant petty bourgeoisie in the major communities.

The process of strengthening the relative position of the petty bourgeoisie would at first glance appear to heighten the potential for overt conflict and to fuel opposition movements in the region, but this is not what happened in the Tehuacán Valley. The reasons are complex, requiring an understanding of community social organization. Alain de Janvry, commenting on the growth of the petty bourgeoisie throughout Latin America, suggests that it serves as a "buffer between the national semiproletarianized and landless peasants: economic interests are tied to those of the bourgeoisie while their ideological identification lies with the bulk of the peasants, who recognize them as political patrons" (de Janvry 1981:253). He goes on to point out that the political stability created by the capitalist development process itself and the economic opportunities created for the petty bourgeoisie are a contradiction that leads to greater proletarianization and accumulation of resources. It is significant that until 1984 unequal distribution of water spawned neither class-based violence nor political mobilization. The reason lies within the structure of interpersonal and class relationships, which are part of daily life in the communities.

Traditionally, the indigenous population in the region had an ideology of resource sharing within the community. As the region became increasingly incorporated into the wage-labor system at the beginning of the twentieth century and then into the capitalist market system during the 1930s, ideology changed to one of individual accumulation. Yet vestiges of the traditional ideology remain. It is commonly held that people of the community should have the first access to irrigation water, whether it is sold or distributed through sharecropping. This provides a degree of legitimacy for unequal water ownership, which in turn minimizes class-based hostility on the part of those who have no water.

INTEGRATING STRUCTURES

Although there is a long history of conflict in the valley, class-based opposition or reform movements have failed to evolve because of the institutions that integrate different segments of the population. The wealthy, indigenous agriculturalists who occupy pivotal positions in the economic and social structure of the region control crucial resources such as water, land, jobs, and, often, credit.

Waterlords and Distribution of Water

In each of the large valley communities there are families that own water rights worth thousands of dollars. These families accumulated those rights in different ways. Some inherited water rights that were bought when the haciendas were divided. Others invested in the construction of new *galerías* that were successful. In at least two cases, moneylenders built fortunes by foreclosing on peasants who had small parcels of land with water rights attached; over time, multiple foreclosures led to significant accumulation.

Few families that control large amounts of water are able to use it all. In addition, there are people who own water but do not live in the valley. Although their origins are in the valley communities, they now live in Mexico City, Tehuacán, and Veracruz. For them, water is an investment, and they must develop ways to ensure a return on their capital.

Individuals who control large amounts of water may pursue a variety of sharecropping strategies simultaneously. The system is an effective means of appropriating surplus and creating a social network in which they control the critical resources. On the other hand, the agriculturalist who does not own sufficient water needs to create a dependable relationship with someone who does. In order to gain access to water, an individual is often willing to become part of a quasi-hierarchical network and to establish a patron-client relationship. Although there is a definite dimension of subordination in the relationship, the relationships are perceived, both publicly and privately, as evolving from friendship or kinship bonds. Traditionally, the act of keeping the resources within the community has legitimized inequitable ownership of water. As the economy is increasingly incorporated into markets, community acceptance and legitimization is less and less important.

Patron-client ties between two people of unequal status are played out in many different ways. The relationship includes a whole range of obligations and expectations, especially if it is formalized through *compadrazgo,* a godparent relationship. Few people use the term "patron" to refer to an individual with whom they have a dependent relationship, but social relationships in the valley are laced with dyadic patterns of friendship that transcend class lines and provide the integrating structures of the community.

Sharecropping

Sharecropping is a mechanism for controlled distribution of water in some valley communities. For those who depend on sharecropping for access to water, creating a dependable patron-client relationship with a person who owns surplus water is crucial.

There are various types of sharecropping in the valley. The most complex form involves three people: one person provides seed, oxen, equipment, and most of the labor; the second provides the water; and the third provides the land and contributes some labor. This pattern is arranged by either the landowner or the water owner. The crop is split in thirds, with the parties marketing the crop on their own or using it for domestic consumption. The latter is particularly common when the crop is corn. If the corn is harvested as *elote* (sweet corn), it may be sold in the fields to marketers from Chilac, Zinacatepec, or Altepexi, who in turn sell it at the wholesale markets in Mexico City. Tomatoes may be sold to buyers who take the produce to urban markets, or sell it in the market in Tehuacán. When crops are sold in Tehuacán, the cost and earnings are divided equally between the three parties.

An individual who has land but no water can get water directly by sharecropping. The person with the land provides the seed, fertilizer, oxen, and labor; the second party provides the water. The crop is divided equally. Most commonly, water owners who enter this type of relationship either do not own land (because they have nonagricultural occupations and bought the water as an investment) or own very little land.

The final type of sharecropping occurs when an individual has water and land but does not want to do the agricultural work. This pattern evolves most commonly when a man is too old to work or when a widow is left with land, and there is no one in the household who can work in the fields. In exchange for a quarter of the crop, workers (often relatives) are hired to cultivate the land and harvest the crop. Some of the largest shareholders in irrigation associations are older women who do not work in the fields themselves but adroitly coordinate production activities.

An example of this pattern is María, one of the largest owners of water in the region. Eight years ago she was selling some of her water regularly but had also established sharecropping relationships with eight nephews and a cousin. The ownership of water had been accumulated by her husband, who was a moneylender and trader. He had invested his profits in new *galerías*. Nevertheless, most of his water rights were acquired when people did not pay their debts.

At first, sharecropping was maintained within the extended family, where María eventually became the dominant figure. But over a ten-year period she shifted her strategies by selling a larger part of her water and sharecropping to only one individual, a distant relative; she claimed that family and business did not mix. Her relatives were reluctant to pay regularly, because she was family and could afford to wait. In contrast, her new sharecropper was very reliable.

Sharecropping, although it is part of the capitalist mode of production, plays an important role in integrating different segments of the population and perpetuating the status quo. In some cases there is a sense of social responsibility, rights, and privileges based on kin or fictive kin ties. But the frequency of sharecropping is declining because there are more efficient ways to maximize profits. Increasingly, where sharecropping still exists, it is temporary and strictly a business transaction.

RENTING AND SALE OF WATER

Less personal ways of earning money from water have evolved in several of the major communities of the valley. Water can be rented to one or more individuals in long-term agreements, or sold. Renting constitutes a transfer of a specific amount of water for an extended period of time at a given price. The sale of water, in contrast, is for a certain amount of water. These types of water transactions are different from the direct sale of water rights, rights that give the owner long-term control over a particular quantity of water produced in a spring or well.

When families move from the valley and anticipate a long absence, they often rent out their water and land. These transactions are written but seldom certified by a lawyer or public official. Although most renting takes place between extended family members or friends, a new pattern is evolving: renting water and land to powerful families living in the city of Tehuacán. Both private land owners and *ejidatarios* have recently rented land to the Tehuacán-based families, which are in turn forging a new form of neolatifundism in the region.

The water market of San Gabriel Chilac is a major institution created for the exchange of water. It is used by campesinos who need to purchase water for a specific crop, individuals who temporarily have a surplus of water, and those who prefer to sell water rather than worry about sharecropping or renting problems. There are a number of factors that create temporary surpluses, the most important being the size of family lands cultivated at a specific point in time, the types of crops being grown, the stage in the planting schedule, soil type, time of year, and micro-climatic variations.

The water market developed in Chilac because the community is located at a higher elevation in the valley than most of the other major communities, and water can easily be sent to fields located at lower elevations, in effect making it a regional market. Another factor that contributed to its growth was that many people in the community have small parcels of land with rights to small portions of water. These holdings represent fragments of land and water that were once given by the community to each household, but that have diminished over time through inheritance. Although this water does not enable individuals to irrigate all of their crops, households can irrigate a hectare or two once a year by exchange and sale of water. The market provides an impersonal way to facilitate these transactions.

Water sales are held at a daily market in Chilac. During the week water is bought and sold between nine and eleven in the morning; on Sunday the market is shifted to the afternoon. Between 50 and 120 people gather to buy and sell the rights to use water from the valley springs and chain wells. The water is measured in units of water flowing from a specific channel or spring. Although most of the buyers are from Chilac, buyers also come from the neighboring communities of San Mateo, Altepexi, and particularly San José Miahuatlán.

People selling water announce the source of their water, specifying the name of the spring or canal, and give the day and hours the water will be available and the price they are asking. Although primarily men attend the market, women may participate as well. Interested people gather around to bargain and to discuss weather and crop conditions. Potential buyers move from one seller to another. Direct trading of water rights, although not the most common form of transaction, may occur, but is perceived as dangerous because of price fluctuations, which have in the past led to conflicts between individuals and families. Depending on the season, rights to water may be sold several weeks or even months in advance.

The majority of the buyers are peasant-farmers who need the water for their crops of corn, garlic, and tomatoes. Some people speculate, hoping to buy water inexpensively and to sell it at profit, but this is uncommon except for the *apisadores* or *canaleros*—water brokers—who play a key role in the water distribution process. If people have water they cannot use or if they need cash, they can either sell the water directly to a broker or commission a broker to sell the water. Some people handle the transactions without a broker, but are subject to pressure from friends and relatives, which often results in lower prices. Although brokers may work for anyone, most have clients who own rights to large amounts of water. Brokers may speculate with water, buying inexpensively one or two months before it is scheduled to be used, and selling it at a profit if the demand is high; brokers provide the same services for clients.

The original role of the broker is to distribute the water, much like the *aguador* in neighboring communities. The work entails opening and closing dams that control the flow of water down the canals to the fields. Campesinos or their laborers make small dikes to channel the water into the fields. Flow in the canals is tabulated by measuring with sticks lodged in the canal's side to mark the depth of flow belonging to each owner. Despite periodic conflicts over measurement and timing of irrigations, the brokers have played a key role in a highly efficient system of water distribution.

Case 1

Ricardo is one of ten water brokers in Chilac. He keeps careful records of all of his transactions in his omnipresent shirt-pocket notebook. After he makes an agreement with a client, he gives the client a piece of paper (*cédula*) that has the details of the source and quantity of water, the number of hours of water, and the date the water will be available. Like the other brokers, he deals primarily with men, but some of his most important clients are women. He is always speculating on water, because many people come to him on short notice desperate for water. Ricardo travels around the region by motorcycle. Trusted by all of his clients, he is paid for every hour he supervises and receives a commission for water he sells.

The marketing system uses the dynamic of supply and demand to distribute a constant flow of water for the changing needs of the agricultural community. Because those most successful in the capitalist mode of production can pay the highest price for water, families involved in subsistence production must obtain water in other ways. The social cost for the economically "efficient" use of water is increased proletarianization of subsistence producers who do not have water of their own or social relationships to supply them with water.

Case 2

Pedro and Carmen live on a small lot in the back of Carmen's parents' compound. They have six children, four of whom are under ten. For eleven years Pedro sharecropped with a wealthy waterlord who had accumulated rights to over one hundred hours of water per month. The man had obtained these rights by lending money to campesinos who owned small pieces of land, half acres or less with water rights that they had inherited. When the campesinos could not pay their debts, he foreclosed, taking the water rights that they had used for collateral. The waterlord then allocated part of the water through sharecropping and sold the rest at the water market. When he died, his widow decided to sell all of the water through a broker at the water market and not sharecrop. Pedro lost his dependable access to water.

Pedro did not have much land of his own, one-half acre divided into three parcels. Three years in a row he had to borrow money to buy water to irrigate his fields. He planted a commercial crop, garlic, instead of a subsistence crop. The garlic required frequent irrigation. During the dry season water was very expensive and he had to borrow money to pay for the water. The price of garlic was low two years in a row, and Pedro was forced to sell his land to pay for his loans. In contrast to some other households in the community, none of Pedro's extended kin had water, land, or jobs they could offer him. Not wanting to work as a *peón* in the valley, Pedro moved to Mexico City to work with a brother who helped him find a temporary job.

The Water Judge and Trading of Water

In Ajalpan another institution has been developed to distribute water. A person with extra money may tell a *juez de aguas* or *aguador* (water judge) the time and the number of hours of water that he or she would like to trade. The judge then looks for a person who will use the hours of water on the promise of returning the lent water at a specific date. The *aguador* is not paid for the service, but gets a small gift or a tip. Elected by the water owners, the *aguador* is in charge of water distribution for the whole community. Although the *aguador* plays an important role, most forms of water exchange are direct transactions between individuals who handle the selling, exchange, or sharecropping agreements themselves.

DECENTRALIZATION

The institutions developed to distribute water in the Tehuacán Valley reflect the decentralization of the irrigation system. The decentralized system served sectors of the valley that were involved in a subsistence mode of production by developing mechanisms for exchange of water between unrelated households. As the region became increasingly incorporated in the capitalist mode of production, new mechanisms evolved for distributing water that facilitated the accumulation of capital and water resources by one sector of the population. The water market in Chilac, with its brokers and water futures, is the most fully developed water distribution institution of this nature in the valley. In contrast to sharecropping, which is also embedded in capitalist production, the market contains no ideology of ongoing social relationships, however hierarchical. Markets integrate different segments of the population and perpetuate an acceptance of the status quo.

There is growing frustration as campesinos with limited water resources struggle to survive in the valley. Without money to invest in wells or *galerías* they have witnessed the rapid rise of prices. The growing wealth of those who own water rights within and outside of the communities is obvious to everyone. The process runs against the traditional concern of social good and justice that is ingrained in Mexican water law. Although most campesinos are not fully aware of Mexican water law, they do know that the system of water distribution in Tehuacán is different from that in many other parts of Mexico.

Part of the structure of dominance is reflected in the absence of organizations that make it possible for the campesinos to participate in the decision-making process. In many parts of Mexico the Comité Regional of the Secretaría de Agricultura y Recursos Hidráulicos includes representatives of campesino organizations. In the Tehuacán Valley, there is no campesino representation. The Confederación Nacional Campesina is weak in the region and is controlled by a network of caciques who benefit from the private ownership of water. The organization has never made an effort to support nationalization of the resource. There are no alternative unions that could mobilize the campesinos, and the two dominant political parties of the region, PRI and PAN, are controlled by individuals who support the status quo and who in many cases own considerable quantities of water.

The structure of exclusion is an important aspect of the power of the cacique network that controls the region and leaves the campesinos without access to water. The fact that the campesinos are excluded from the decision-making process is a dimension of unequal power (Gaventa 1980:11). People are well aware of their relative powerlessness. In the opinion of some, the federal government, at the national and local levels, is controlled by self-serving figures. Furthermore, they assume that any federal effort for reform would be co-opted by regional caciques, who work with federal officials. According to many, giving control of water resources to groups or individuals outside the community would make water much more vulnerable to manipulation by powerful mestizo families based in Tehuacán or by politically powerful agriculturalists in the valley; therefore, the situation is better unchanged.

SHARED IDENTITY AND COUNTERVAILING ALIGNMENTS

Shared identity is one of the major forces unifying the indigenous population of the Tehuacán Valley, despite its stratification. The identity has been forged by persistent threats and attacks from people valley inhabitants perceive to be outsiders. The Spanish, and later the mestizos, stole their land and water. Surrounded by mestizo-owned haciendas, mestizo or Spanish merchants settled in communities during the late nineteenth century. They were part of a social and economic network based in the city of Tehuacán. Despite extensive trade relationships between the Indians and the Spanish merchant families, the groups regarded one another with suspicion and rarely intermarried. During the revolutionary period, most of the Spanish and mestizo families moved to the city of Tehuacán for their own protection. Their stores and warehouses were often burned and looted. Once the Revolution was over, the mestizos did not return to the Indian towns. During the 1930s the haciendas were broken up and the indigenous population regained control of much of the valley. But during the 1960s mestizos based in Tehuacán again began to accumulate land and water resources in the valley to establish agribusinesses.

Neolatifundism

The emergence of neolatifundias, capital-intensive agribusiness, in the valley is perceived by many campesinos as the major threat to their access to water and land. Neolatifundias are a part of modern agriculture in many regions of Mexico where private entrepreneurs have managed to accumulate significant land and water resources. With access to private and public credit, the entrepreneurs have often created highly sophisticated systems of production whose growth may be at the expense of the campesinos of the region (Warman 1976:66).

In an examination of recent trends, David Barkin and Billie R. DeWalt argue that "the modernization of agriculture through improved technology and the application of the theory of comparative advantage have done little to resolve the problems of rural development or eliminate hunger in Mexico" (1988:52). They go on to say that government policies have encouraged the production of commodities for the upper and middle classes, which is not real economic development for the majority of Mexicans (1988:52–53).

The neolatifundism that has emerged in the Tehuacán Valley is primarily aviculture, raising poultry for meat and eggs. This form of agriculture is highly capital-intensive and technologically sophisticated. The Mexican poultry industry is eleventh in the world in terms of production, and the Tehuacán Valley has emerged as one of the centers of this industry (Agribusiness Associates n.d.:48). The poultry farms link the valley to international agribusiness. Most poultry growers buy processed feed from Purina and veterinary supplies from major international pharmaceutical companies.

Although there are more than ten major poultry operations in the Tehuacán Valley, the industry is dominated by one family whose company has the second-highest total sales of eggs in the country. In addition to eggs, the company is one of the leaders in poultry sales (Agribusiness Associates n.d.:55). This farm complex is vertically integrated with interlocking subsections officially owned by different family members and strategically located in different parts of the valley. This highly modern operation includes laboratories that manufacture some medication and vaccines and a processing plant that uses fish caught by company ships and grain harvested from company fields. The company has its own hatcheries and a transportation subsidiary that delivers chickens and eggs throughout the country. In addition to the poultry operation, the company operates cattle and sheep farms, vineyards, and orchards. Large amounts of water are necessary for all of these operations. Part of the water supply is groundwater pumped to the surface by pumping stations. In other cases, the company has been able to purchase major springs, some of which are located near valley communities. Particularly striking is a spring that surfaces next to San Marcos Necoxtla. Although the SARH office had detailed records of campesino water ownership, the files we checked on the water ownership of the largest farm owners were empty.

Employment in valley agribusinesses has fluctuated and is mechanized. It is estimated that the major poultry business complex employs over five thousand in the Tehuacán Valley, many of whom are paid below the minimum wage (*El Día* 1977:2). Most workers do not receive social services. In an effort to avoid unionization, rural workers are placed on farms located as far away from their home community as possible and frequently moved from job to job. The effort to prevent unionization has been successful.

The expansion of the neolatifundias in the valley has had an impact on the resource base. Not only have the owners of the farms been able to get pumping rights at a time when the expansion of the *galerías* is prohibited, but they also have bought or rented land throughout the valley. Although pumping is lowering the water table, new complexes are being built. When land comes up for sale, the neolatifundia owners have the capital to bid for it. The pressure has increased prices for both land and water, often making them too expensive for most campesinos to purchase.

The farms and poultry-processing operations have created jobs in the region. Young people from the valley communities do work as laborers on the farms and at the urban-based plants. As a result, the urban-based families that have purchased the campesinos' means of production are now hiring and exploiting their labor. The process has reinforced a we-they perspective in the indigenous communities.

The class of urban-based *neolatifundistas* dominates the regional power structure. They have the power and wealth to appropriate land

and water, influence the selection of regional PRI candidates, and pressure bureaucratic leaders. As part of their power, they have built patronage networks linking them to valley communities, thereby defusing potential overt hostility directed at their holdings.

We have already discussed competition for resources in the valley, which has historically pitted community against community and groups within the same community against one another. But the mestizo elite based in Tehuacán is regarded as the greatest threat by indigenous families. For most campesinos, it is the ominous shadow of this power that shapes their fear of losing control of land and water and as such has forged a fresh identity and tarnished internal class sensitivities.

As a result, the threat from the outside deflects people's attention from class conflicts within the communities. The ideology of unity is strong and often reinforced in speeches and conversation. It is often pointed out that they are all campesinos (people of the country whose life is tied in one way or another to agriculture). Equally important, they regard themselves as Indian. The Indian identity is important, although in many ways they are not culturally different from non-Indian Mexicans. Judith Friedlander, referring to another region in Mexico, suggests that "from the perspective of culture the Indian villagers are virtually indistinguishable from non-Indian Mexicans. Where they are different is in their social status within the larger society" (Friedlander 1975:xv). She goes on to say that the people are eager to lose their Indian identity and that it is the outsiders who want them to be Indian (1975:182).

This does not seem to be the case of the indigenous population of the Tehuacán Valley. The history they recount is one of pride and unwillingness to accept subordination. Regardless of social class position, they are proud that they, Indians, control the valley. They are proud of their historical roots, their ability to speak their own language, and their struggle for control of the land and water. It is this perspective that gives the population a consciousness of unity, despite internal conflicts. This is the web that binds the community elite to the semiproletarianized and landless campesinos. It is also the ideological component of the structure of power, contributing to the identification of the landless and waterless campesinos with the waterlords of their community.

8. CONCLUSIONS: THE STATE VERSUS LOCAL INTERESTS

THE HISTORY of agriculture and irrigation in the lower Tehuacán Valley is long and complex. Although many calcified canals lie abandoned, testimonials to the labor of previous generations, a key to understanding the valley is still the social organization of water control and distribution. As the history presented in this volume documents, the competition for water and the cooperation required by irrigation are forces that have continually divided and unified valley populations.

In contrast to the major agricultural regions of Mexico, where water resources are owned and administered by the federal government, water is privately owned in the Tehuacán Valley. The importance of private ownership has traditionally been equated by the people of the valley with local control. For most of them the state represents institutions controlled by outsiders who are generally regarded as a threat. At the same time, valley agriculturalists are frustrated by their inability to capture state resources.

The central government is seen as both a potential resource and a threat. There is a generally shared feeling that state leaders do not have the best interests of the peasantry, the landless, or the indigenous population of the country at heart. At the same time, people are aware that some resources are being allocated to rural health and development projects. The fact that the Tehuacán Valley has not received what many regard as a fair share is blamed on both national and regional leadership. Yet all people in the region do not share the same desire for greater federal/state involvement in the region.

The large communities of the valley are highly stratified, a process accelerated by the integration of the region into the capitalist economy and by the expansion of agricultural production for urban markets. Rural workers without land or water resources have very different expectations of the state from those of the people who control significant amounts of land and water; yet until recently they have shared a distrust of state-generated reform.

One of the most significant postrevolutionary events was the implementation of the long-awaited agrarian reform. In the introduction we stated that the organization of power and interests at the regional and community levels is a critical variable influencing how state policies are interpreted and enacted at the local level. In the cases examined in this book, the Mexican agrarian reform was implemented differently in different areas of the same region, because of the constellations of power at the community level and the way in which they articulated with the national centers of decision making. Although the expropriation and redistribution of land occurred in most parts of the valley, water was often not allocated with the land, with far-reaching consequences for future generations.

In the highly stratified community of Chilac, the *ejidatarios* are still very poor and politically uninfluential. Although they failed to get the rights to the water previously used to irrigate the lands, others were able to do so by breaking the laws and manipulating surveys to allow the resale of water. The result reinforced basic class differences between those in Chilac who controlled the means of production and those who did not.

In contrast, in Altepexi the majority of the agriculturalists became part of a single ejido organization formed in the late 1930s. As in Chilac, water was not redistributed with the land. A small number of individuals gained ownership of a relatively small amount of water, which they used to irrigate a fraction of the total ejido grant. The important difference was that in Altepexi ejido members who had some water formed groups with other *ejidatarios* without water to raise capital for constructing *galería* systems. As a result, there was an important linkage between the ejido organizations and the irrigation associations. In contrast, *ejidatarios* in Chilac were marginal to the individuals who controlled water and did not have the resources to make significant contributions to the development of new *galería* systems. Ultimately, however, the process contributed to the process of social differentiation within the ejido as well as within the community, strengthening the position of a small number of families.

In Chilac and Altepexi the implementation of the federal land reform program was modified by people who had the power and desire to do so at the regional or community level. State power was manipulated at both the national and local levels. The haciendas of the valley were broken up, but many who had hoped to benefit from the reallocation of resources were left without water and were unable to cultivate. Equally important, they had a new reason to distrust the state.

Despite campesino disillusionment with the agrarian reform, state leaders did not attempt to control water resources, as was their right. Campesinos had access to land and knowledge of a technology uniquely appropriate to the local environment and requiring no expensive machinery: the *galería* system. The successful completion and management of over one hundred *galería* systems without any government help represents an immense achievement by the agriculturalists of the Tehuacán Valley. The nature and evolution of this system has been carefully described in this book.

The collective organizations created to finance and manage the construction and operation of *galerías* kept water ownership within the valley. Local resources and skills were used for constructing an energy-efficient system. Although control of the system remained local, the markets for the produce were national, because the water was used for cash crops, not only for subsistence production.

Many agriculturalists participated in the creation of *galería* associations. Given the capitalist nature of the investment and ownership rights, the system has been much more equitable than if only the most wealthy agriculturalists had drilled wells powered by electric pumps, as has happened in other parts of Mexico. Nevertheless, this process combined with the open water markets has led to unequal control of water and has accelerated social differentiation in the region. In a capitalist economy, collective organizations such as the *galería* associations do not generate equity.

For those who gained ownership over water through the building of the *galería* systems, the latter came to represent a sacred resource. The ownership of shares in the system has been a critical factor in maintaining and generating differences in wealth in the valley. In Ajalpan, a neighboring community to Chilac and Altepexi, Luis Emilio Henao found that a number of associations were dominated by individuals who owned more than 50 percent of the shares, did not cultivate their land, sold their water, and sharecropped (Henao 1980:150). A similar pattern evolved in Chilac. The pattern is not as clear in Altepexi. Yet in all three cases, wealth gained from water ownership has been important as families diversified their economic activities by buying land, trucks, or urban housing.

Despite the fact that the Tehuacán Valley is located in one of the major federal irrigation districts, the Papaloapan District, the federal government has maintained a policy of nonintervention. There are various explanations for this policy. First, according to officials, the *galerías* are products of peasant labor and investment. Thus, it would be unfair to take control of the privately developed resources. Second, any effort to nationalize the systems would generate great social unrest in the region. Third, the system is being managed efficiently. Fourth, (but not a reason given by SARH officials) wealthy, influential people in Tehuacán are making a considerable amount of money on some of the wells in the valley.

Since internal administration, allocation, maintenance, and conflict resolution have been left entirely to the associations, elite *socios* and their supporters have had considerable autonomy to implement their programs. The decentralized nature of the irrigation system in Tehuacán has generated a series of conflicts within the collective organization, between associations, and between communities. Many of the conflicts have festered and remained unresolved over long periods of time. In some cases, the federal government has been called in, but as in the case between the two associations described in Chapter 6, it has been reluctant to interfere with established property ownership and legalized water division. Ownership of water is based on customary law of the region and enforced by popular support. This support is based on the internal distributive system described in this book. Conflicts are usually resolved locally, because participants do not want to bring in the federal government, which might restructure the total pattern of ownership. Government regulation could have encouraged cooperative maintenance, but government officials refused to get involved, despite requests from several of the associations.

The Tehuacán case illustrates several important dimensions of state penetration. Leaders of local-level bureaucracies may have tremendous latitude in how they use their derivative power. They are often key brokers linking communities to national institutions. The nature of the relationships between these local-level bureaucratic heads and different segments of the communities is a crucial variable affecting how policies are implemented.

The organization of power and interests at the regional and community levels is a critical variable that influences how state policy is interpreted and enacted at the local level. These interests may be divergent and often in competition with one another. Although state leaders may be able to co-opt local officials, local influentials can also co-opt state appointees for specific interests. This is particularly important if the state gives local leaders the power to define and implement programs. Even though water is officially nationalized in Mexico, in Tehuacán the state allocated the control of water to whoever had the capital to develop water resources. Land, on the other hand, was redistributed by the state, but it was worthless without water. In contrast, once water was developed, it became a marketable commodity.

In areas where the state is expanding its power and role in the countryside, there is often an inherent conflict between the pressures for centralization and the desire of many local residents to retain autonomy. In the Tehuacán Valley, resistance to state penetration has been framed in an ideology that emphasized ethnic unity over intercommunity class differences and in a system of water allocation that incorporates a significant number of agriculturalists. It has also been based on the historically accurate belief that the indigenous population of the valley does not control state institutions. If water resources were taken over by the state, urban-based regional elites would eventually gain control of them.

The water resources of the valley are finite, and the increased digging of wells has generated a growing number of conflicts over water. As these conflicts escalate, there is new pressure on the government to play a more active role in the management and distribution of water. A second and more important process militating for greater government involvement has come from the increased mobilization of the waterless agriculturalists of the valley. Declining real wages, efforts to increase agricultural productivity, growing unequal access to water, and population growth have increased pressure on water resources. Agriculturalists who own no water or only limited amounts are denied access to water through traditional means, forcing many of them out of agriculture. As a result, the local alliances mobilized to maintain regional autonomy on water issues, which had transcended significant local-level factionalism and class awareness, have begun to disintegrate.

In other words, the ideological underpinnings for local autonomy in resource control based on ethnicity and a shared view of the state break down when the pressure on the resource is so great that it cannot be distributed in such a way that it is perceived by those who do not control it as legitimately accessible. At this point, class awareness may become a basis of political action.

The unequal distribution of water has emerged as a major political issue in the mid-1980s after being a politically taboo subject. In Chapter 5 we discussed factors that inhibited the overt expression of class awareness and conflict between those who controlled water resources and those who did not. In the mid-1980s, PSUM (Unified Socialist Party of Mexico), a leftist political party, emerged as an important political force in the valley. PSUM leaders helped mobilize poor peasants and articulated long-held concerns. The growing disparity in wealth and in control of means of production, Mexico's economic crisis, and the Mexican government's loss of legitimacy in the eyes of the peasants of the valley have led to a new political expression. The government's loss of legitimacy can be traced to the campesinos' frustration at not receiving new government services, the growing rate of inflation, and irritation with government corruption.

The conflict over water underlies an important class-based division that common ethnic bonds will not soften. For a significant segment of the population, inequitable control over basic resources is the most pressing problem facing the valley. No longer is the fear of an outsider takeover of water resources as strong as the frustration of being marginalized from their means of livelihood. Nationalization of water is seen as a step toward gaining access to water.

Agriculturalists who do own shares in water associations remain adamant in opposing nationalization and centralization of management. In other parts of the world, centralization of local water systems has led to both inferior maintenance and less effective use of water (De Yong 1981). In areas of Mexico where the state controls water and the distribution process, water prices are often subsidized, and efforts are made, although not always successfully, to maintain equity in access. However, in order to gain access to water, growers must get permits from state officials; as a result, the state has increased control over who can plant what crops, where the crops can be sold, and at what prices (Whiteford and Montgomery 1985). Yet in other cases where formal centralization of authority has occurred, local control has been maintained. In instances in which local cooperative organizations were already strong and retained control over the political process, the state has let the local groups administer their own affairs (Maass and Anderson 1978: 366–367).

As we have shown, the power of the indigenous elite is limited. More powerful families based in Tehuacán and with ties to the centers of political power in the Mexican state dominate the valley. It is the interplay between the regional mestizo elite, the indigenous elite, and the government that will eventually determine the fate of indigenous control of water and land resources in the valley.

At no point in the history of the valley has a whole community or population rallied to a particular set of symbols to produce a unified political or economic front. Conflicts over resources have occurred between indigenous communities as well as between groups with different class positions. Yet the people of the large communities such as Chilac and Altepexi have shared an identity based on language, customs, and patterns of interaction which they feel distinguishes them from urban-based mestizos. Growing population, increased pressure on resources, greater economic diversification, and proletarianization have generated new and greater internal divisions within the population.

During the 1980s new irrigation associations have been organized. Nine major agroindustrial cooperatives, including twenty-five new irrigation associations, have been developed with government support. It is projected that water from these new wells will increase the land under cultivation by 34 percent. The majority of the new cooperative members are also *ejidatarios* and therefore belong to local ejido organizations. There have been heated negotiations between cooperatives, ejidos, and communities over digging rights for wells and canal right-of-way. Many of the most powerful families of the valley are involved in the process, but some feel threatened by the new organization. The government has played an increasingly important role in negotiating these conflicts.

The long history of cooperation and competition for water resources in the Tehuacán Valley has entered a new stage. The towering mountains and deep *barrancas* that have given the valley its form still stand, but the currents of change are sweeping the ancient valley floor.

NOTES

1. *Mexican Rural Development and Irrigation*

1. The emergence of the state as a dominant force in people's cosmology has received little analysis. The Gudi of Ethiopia, like many people throughout the world, have symbolically transformed state penetration of their domain. "The spirit group meetings are a transformation of the proceedings of the national courts that have been instrumental in exploiting the Gudi. . . . New spirits appear unpredictably and reveal new powers that can affect people's destiny" (Bauer and Hinnant 1980:230).

2. In Mexico the term *campesino* describes a person who earns a living based on agricultural activities. We will use *campesino* whenever possible instead of *peasant* in this text. The term *peasant* lumps the majority of Third World rural inhabitants into a single category, thus implying homogeneity not only in culture but in role structure. Rural life is much too complex in terms of roles, status, and ethnicity to lend any scientific significance to the category of "peasant." As we will show in the chapters that follow, the rural population in the Tehuacán Valley takes part in economic, social, and political activities ranging from landless wage laborers to wealthy water owners holding high political office. Some individuals are wage laborers some of the time, farmers on their own land, and skilled basket makers at other times. The time spent participating in different activities varies over relatively short cultivation cycles and even more during an individual's lifetime. These multiple roles, and associated individual behavior, cannot be subjected to the reductionism implied in calling such persons with their role options "peasants."

We do, however, recognize the vast literature on "peasants," to which this work relates, and we have kept the term in a few instances for comparative purposes.

3. Because of their interest in "smaller" and "localized" irrigation systems, many anthropologists and other investigators have had little concern

with or involvement in the controversy over the hydraulic hypotheses. The few studies that have been done have examined irrigation organization and activities on the local level and the state's involvement, if any, with local affairs. Millon, Hall, and Díaz (1962), in a comparative study of seven "small" irrigation systems, concluded that there is no clear relationship between the degree of centralized authority, the size of the irrigation system, and the number of persons it supports. Their conclusions were based on data from seven irrigation societies that had been studied empirically and analyzed objectively. Two involved single-village irrigation systems: the Sonjo in Kenya (Gray 1963) and Pul Eliya in Sri Lanka (Leach 1961). Four involved multicommunity systems: the Twelve Village System in Japan (Eyre 1955), El Shabana in Southern Iraq (Fernea 1959), the Nahid in South Arabia (Hartley 1961), and Teotihuacán in Mexico (Millon, Hall, and Díaz 1962). The seventh, Bali (Geertz 1959), was not classified. The populations served by the seven systems varied from 146 to 40,000, and the area irrigated ranged from 135 to 13,600 acres. Millon (1962:80) concluded that there appears to be no relationship between centralization of authority over water allocation, the number of people involved, and the size of the system. In other words, both large and small systems have existed under centralized control.

4. Eva and Robert Hunt (1976) reexamined Millon's data and made the assumption that "centralization of allocation" meant the highest level where allocation decisions were made. They agreed on the Sonjo, Teotihuacán, and El Shabana; were unable to interpret Nahid; disagreed on Pul Eliya and Bali; and used their own data from San Juan, Mexico, as a further case of centralized control. The Hunts pointed out that in Pul Eliya, irrigation was the most important factor in local social organization. Also, the central government was very much involved in the regulation of legal land tenure, allocation of irrigation water, and major repairs of the village irrigation tanks. From this they concluded that the system could not be classified as decentralized. In the case of Pul Eliya, it was bureaucracy that acquired and organized labor to build the tank systems, and with the passage of time and many changes in government, the local irrigators appear to have allocated water with complete autonomy. Under such "normal" conditions, the system appeared decentralized, but government power and its potential for legitimate use by intervention in local affairs continued to exist. Clearly, "centralization," in order to be a useful concept, must be more carefully examined and defined.

5. Eva and Robert Hunt (1974), in a study of irrigation in San Juan, Oaxaca, formulate the hypothesis that centralization of authority (later to be called unification) over irrigation is developed to reduce conflict under conditions of scarcity produced by pressure on land and water resources. They go on to say that centralization is not adaptive under all conditions, but it is most effective just within the local territorial extension of the irrigation system. Centralization consists of authority in a local political and economic hierarchy articulated with national government institutions. The administration of irrigation and conflict resolution is a major concern for elites with extralocal connections.

2. *The Tehuacán Valley*

1. Finding accurate and consistent data on land area and land use in the state of Puebla is very difficult. Sources are contradictory and vary tremendously. The third, fourth, and agricultural, livestock and ejido censuses show that in 1950 land for agriculture and forest measured 144,231 hectares; in 1960 it was 149,849; and in 1970, it had been reduced to 80,216 hectares (SARH 1982). The figures published by the Secretaría de Programación y Presupuesto differ from those of SARH, which, again, differ from the figures provided by the local office of the Papaloapan Commission in Tehuacán. Wherever possible, we have used data provided by the Papaloapan personnel who have supervised the surveys and made the calculations.

2. Scholars have defined regions in a variety of ways: by geographical characteristics, political divisions, trade domains, and cultural boundaries. Leon Veerman confined his study to the former district of Tehuacán, which includes the *municipios* of Ajalpan, Altepexi, Cañada, Chaltepec, Coyomeapan, Coxcatlán, Chapulco, Chilac, San José Miahuatlán, Eloxochitlán, Nicolás Bravo, Tepanco, Tlacotepec, San Sebastián, Vicente Guerrero, Zapotitlán, Zinacatepec, and Zoquitlán. The SARH study of the Tehuacán Valley includes *municipios* both north and south of the city of Tehuacán, including Ajalpan, Altepexi, Cañada, Chapulco, San Gabriel Chilac, San José Miahuatlán, Santiago Miahuatlán, Tehuacán, Tepanco de López, and Zinacatepec, but excludes Coxcatlán. Our study deals exclusively with the southern end of the valley, beginning with the *municipios* of Chilac, Altepexi, and Ajalpan. We worked extensively in the *municipios* of Chilac and Altepexi, but this overview includes data gathered by other scholars who were part of Scott Whiteford's project and who resided and worked in other *municipios*, including San José Miahuatlán, Ajalpan, and Coxcatlán.

3. The *municipio* is the smallest organizational unit in the Mexican administrative hierarchy. *Municipios* usually include at least one principal population center, one or more smaller communities, and all the surrounding land. The dividing lines between *municipios* have been surveyed and are clearly marked with cement and metal posts.

4. There are 4,451 private holdings, but 3,809 of them are smaller than five hectares. Although we lack precise figures for the valley, it is a common phenomenon in Mexico for people who own private property to also be members of ejidos.

5. The figures on land tenure have to be regarded with a degree of caution. Agriculturalists are reluctant to admit the precise amount of land and water they own. They often give incorrect figures on a census or questionnaire in an effort to finish the interview quickly or because they are afraid of excessive taxation. In addition, few calculate their holdings in terms of hectares. The most common measurement used is the *pantle*, an inexact term used to define a unit of land used for irrigation.

6. Irrigation systems that rely on groundwater must take into account the uneven distribution of water beneath the surface. Variation in composition

and density of subsurface materials results in marked local differences in the depth and yield of aquifers. The following are conditions common to all systems that use groundwater: (1) in arid areas where no perennial rivers flow, groundwater is crucial and is the only practical solution to problems associated with agricultural production; (2) groundwater can usually be obtained near the land where it will be used, which means there is no need for an expensive distribution network; and (3) there is almost always less fluctuation in groundwater supplies than in stream flow (Cantor 1970:43–44).

7. There are many serious and complex problems related to the continued use of groundwater resources. Frequently, the water is present at excessive depths and in inadequate quantities. It may be highly mineralized and may, therefore, be unusable. The water in the Tehuacán Valley is present in large quantities at reasonable depths, but it has a high concentration of minerals and other chemicals, which must be taken into consideration and compensated for technologically if agriculture is to continue.

The most important limitation is that there is only a finite amount of groundwater available in any one area. If extraction exceeds infiltration, the "capital reserve" of water will sooner or later be exhausted. Lowering is most rapid near wells and forms a cone of depression, which is a conical-shaped lowering of the water table. The height of the cone is called a drawdown (Cantor 1970:43). In the case of *galería* tunnels, the drawdown will take the form of a trough with the shape of an inverted V. The drawdown causes the formation of a steeper gradient of the water table near the well or tunnel, which results in increased flow for a short time. The duration of the higher flow will be the time needed for the drawdown to level out. This, in turn, is dependent on local soil composition, hydrostatic pressure, and the quantity of water in the ground. If a number of wells are placed close together, their cones of depression may intersect to produce a general leveling of the water table with no increased flow, because the steep gradient will not be produced (Cantor 1970:43).

8. Engineering and construction are simple and ingenious. First, a preliminary survey of the landscape is conducted by a master (*maestro*), who has two construction options: he may have a vertical well excavated 50 to 100 meters uphill from where it should intersect with the land surface, or he may have a shaft excavated where he feels that the water will be found and then proceed downhill. In the first case, work will proceed uphill and away from the fields. All the vertical shafts are completed before excavation begins on the nearly horizontal underground tunnel. In both cases, the first well is a test well in which the dimensions of the shaft are about 1 meter by 1.5 meters and the depth ranges from a few meters to, in some cases, as much as 70 meters. Depth, therefore, depends on which option the *maestro* has chosen as well as on the highly variable level at which the groundwater aquifers are encountered (Woodbury and Neely 1972:144–145).

The top of the well shaft is generally reinforced with cement, which is raised about 50 centimeters above ground level. This serves to prevent cave-ins at the edges of the well and to prevent excavated soil, which collects in large circular mounds around the well, from falling back in. The diameter of the dirt mound, depending on the depth of the well, commonly measures 20 to 30 meters. This can and does cause problems when the refuse spills onto adjacent farmland. The vertical shafts, called breathing wells (*pozos respirantes*), are excavated at 40-to-60-meter intervals. Apart from their necessity during construction, they function as an access to the *galería* tunnel for periodic maintenance, vital ventilation for the work crews, a passage for the removal of debris, and a means to withdraw water from the system.

All excavation is done by hand with small shovels, metal bars, and sticks. The excavated rock and soil is hoisted to the surface in a straw basket that is a sturdier version of the common Mexican *tenate.* The hoist consists of two Y-shaped pieces of wood across which is placed a log with wooden handles on each end. Twine ropes are used to lower the basket to the bottom, where it is filled with clay, sand, and rocks. The laborers are also transported in and out of the well in this manner; the hoist has no brakes or safety devices.

The *maestro* well diggers have a long tradition of knowing how and exactly where to excavate test wells. By taking into consideration the location of nearby springs, other *galería* systems, and extensive knowledge of the topography and the locations of the aquifers, the *maestro* can specify the location, spacing, and depth of the wells. A string is extended downhill from the first test well, and the intervals for each well are marked on the string. The depth to which each shaft is to be excavated is determined by using a long string, a stone or metal plumb bob, stakes, and a leveling device. The string with the plumb bob is lowered into the test well to the water level, and the depth is marked on the string. This length is then measured out along the horizontal string that goes from the edge of the test well to where the next well is to be dug. The horizontal string is leveled perfectly and the distance from the level string, at the spot where the second well is to be dug, to the ground is the amount the terrain has dropped between the two wells. This distance will be subtracted from the length of the string in the test well, which then is the projected depth for the second well. The procedure is repeated until the depth of the last well is determined before the point where the horizontal tunnel will intersect the surface. These measurements provide a fairly accurate measure of the total amount that has to be excavated and allow the *maestro* to estimate time for excavation as well as cost of construction.

The vertical shafts are dug by work crews of four laborers and a supervisor. Although the supervisor is sometimes the *maestro,* he may be a representative of the organization of agriculturalists that is financing the construction of the *galería* system. The average work day of a six-day week begins at 8:00 A.M. and ends at 5:00 P.M. In the morning, two of the four laborers do the excavating while the other two operate the hoist; after lunch their roles are reversed. The supervisor makes sure that the laborers work at an acceptable pace and that the work is done according to specifications. Depending on the size of the project and the amount of money available, more than one work crew may excavate simultaneously. Most small systems, however, use only one crew.

The final step is the excavation of the nearly horizontal tunnel that connects the bottom of each vertical shaft. It normally begins at the lower end of the line of vertical shafts. The dimensions range from 1 to 5 meters in height and from 1 to 3 meters in width. *Galerías* with high ceilings are usually the result of further excavation of the tunnel floor in order to increase the flow of water. Tunnel sections through loose soil and rock are often shored up with cement or wood beams. The floor of a *galería* with a "normal" flow is covered with water that is about 50 centimeters deep. First, the tunnel that intersects with the surface is made and a surface canal is dug to prevent backup of water or flooding, which can result when the tunnel enters an aquifer.

In order to assure straight excavation and to prevent lateral curving of the tunnel, a set of candles is placed along the floor and the excavators sight along them to the vertical shafts. A series of three or more candles provides an accurate determination of the vertical slope of the tunnel that defines the depth at which the water table is located. Careful determination and control of the tunnel slope limits the amount of water that will enter the tunnel to normal and safe levels.

9. Richard K. Cleek explored the terminology associated with the *galería* system and found some evidence for pre-Conquest origins of some of the wells. The complete Spanish terminology involved in the *galerías* of Tehuacán and elsewhere in Mexico lends weight to the probable post-Conquest introduction of the technique into Mexico from Spain. Some evidence, however, contradicts this view. Most of the *galerías* from Tepeaca, north of Tehuacán, have Spanish names, except some older dry *galerías*, whose sources are not above the dropping water table and whose names are of Nahuatl origin. The practice of Hispanicizing the names of springs, and possible *galerías*, was quite common in Tehuacán, thus destroying any evidence, if such existed, for pre-Columbian *galerías* in the valley. As early as 1601, the Coyoatl and Axoxopam springs were named San Miguelito and San Lorenzo. In the Tehuacán Valley today only five of eighteen pre-Conquest springs have Indian names (Cleek 1972:61).

10. Cleek (1972:15) reports that *galerías* have been built with perforated concrete pipe by the Mexican government to provide water for households in twenty-two dispersed areas of Mexico. Their locations include Baja California, Michoacán, Hidalgo, and Veracruz. Cleek also found that *galerías* were built in Parras, in Northern Mexico, to generate water for irrigation. The community was colonized by Tlaxcalans, some of whom may have been employed as laborers by the Franciscans in the Tehuacán Valley. Cleek suggests that these people learned the technique in Tehuacán and used it in Parras (1972:18).

3. *The Pre-Conquest Development of Agriculture and Irrigation*

1. The Boserup Hypothesis suggests that increased agricultural intensification, in a situation of land scarcity in relation to demands for production, is the result of increasing population pressure through higher reproductive

rates or in-migration (Boserup 1965). If access to productive land does not increase along with population growth, then innovations such as irrigation, orchard culture, and multiseason agriculture are examples of production intensification.

2. The Early Formative period is generally considered to be when there was a movement out of caves to open residence in small wattle-and-daub villages in the center of the valley. The move was necessary in order to reside near agricultural fields, which had become the major sources of food (MacNeish 1972:25).

Synonyms for the Formative period are the Preclassic, Archaic, and Developmental, and the beginning dates vary somewhat, depending on particular sites and their location in Mesoamerica. For Central Mexico and the Maya Lowlands, there is general agreement that 1500 B.C. marks the beginning of the Early Formative or Preclassic (Willey, Ekholm, and Millon 1964).

3. The term "ramage" is used to add specificity to a diffuse, often large, and very heterogeneous network of extended family or kindred that includes all possible cognates. Ramages, as a subset of the kindred, "distinguish the cognate groups which actually do pull together as relatives and which recognize active bonds of identity and obligatory interaction as kinsmen" (Hoebel 1966:370).

4. *Post-Conquest Conflict over Land and Water*

1. The history presented in this chapter is a broad overview. Much more archival research needs to be done on the history of the Tehuacán Valley. Many of the data for this section were gathered from the General Archives of the Nation, the Archives of the Agrarian Reform, and the local archives of San Juan Ajalpan and San Gabriel Chilac. Unfortunately, archives in several towns had been badly damaged by floods or earthquakes. In addition to archival work, oral histories were gathered. Luis Emilio Henao was a member of the research team that worked with Scott in the valley.

More anthropologists are now using longer time perspectives in their research on local politics and the nature of conflict. Mexico, in particular, has been characterized by lengthy disputes over access to land, water, and political power, as exemplified by two recent monographs, one from the Tarascan region (Friedrich 1986) and the other from Oaxaca (Dennis 1987).

2. In one case, a group of Indians moved to Córdoba, Veracruz. They took produce from the valley and remained working in the tobacco fields. They were followed and captured. The mayor of Córdoba petitioned to have them stay in Córdoba and ultimately paid a ransom so that they could remain.

3. By 1932 CROM had organized a large number of unions in the states of Puebla and Oaxaca. Included in the Unionist Federation of Workers and Farmers of the Limited Area of the States of Puebla and Oaxaca were the Workers' and Farmers' Union of the Tilapa Mill, the Workers' and Farmers' Union of the Buenavista Mill, the Union of the City of Tehuacán, the Workers' and Farmers' Union of the Ayotla Oaxaca Mill, the Farmers' Union of Chilac,

the Textile Workers' Union of Xaltepec, the Farmers' Union of the Santa Cruz Ranch, the Farmers' Union of A. Valerio Trujano, the Bottlers' Union of Tehuacán, the All Trades' Union of Tehuacán, the Traders' and Bakers' Union of Ajalpan, the Bakers' Union of Tehuacán City, the Farmers' Union of A. de San Juan Ixcaquistla, the Union of Chilac, the Palm Tree Workers' Union of Ixcaquistla, and the Railroad and Freight Workers' Union.

4. The traditional *jefes* or caciques were usually powerful *hacendados* in the Porfirian administrative hierarchy.

5. Before 1976 there were 12.5 pesos to the dollar. After 1976 there were approximately 25 pesos to the dollar; in 1979 the peso was further devalued to 46 pesos to the dollar, and by the middle of 1983, after tumultuous fluctuation on the international money market, the peso dropped to approximately 150 pesos to the dollar. By 1986 there were 700 pesos to the dollar; in mid-1988, 2,000.

6. The data presented on both Altepexi and Chilac were obtained from *municipio* records and archives in the offices of the *presidentes municipales.*

7. The *caballería* was a variable unit of colonial land measurement. For the purpose of the agrarian reform, it was set at slightly under forty-three hectares.

8. The Hacienda San Francisco, located between Altepexi and Ajalpan, had a total area of 630 hectares, 576 of which were irrigated and 54 of which were pasture. The Hacienda Santa Cruz, located just south of Tehuacán, had a total area of 1,116 hectares. Of this area, 128 hectares were irrigated and 576 were suitable for agriculture if more water were available. The rest was pastureland (*agostadero cerril*). The Hacienda Venta Negra, located next to the Hacienda San Francisco, had a total area of 2,015 hectares: 11 hectares were occupied by residences and other buildings, 70 were for irrigation, 1,413 were for temporal farming, and 521 were pasture. Venta Negra owned the water from Atzompa spring, which was eventually given to the *municipio* of San José Miahuatlán, located about ten kilometers to the southwest of Altepexi. The Hacienda San Andrés, located south of the city of Tehuacán, consisted of 3,442 hectares, of which 44 hectares were for irrigation and 3,398 were pasture. The lands of San Andrés were considered to be of poor quality and unsuitable for agriculture. The Hacienda Buenavista, located to the northeast of Ajalpan, had an extension of 507 hectares: 72 hectares were irrigated, 45 were unirrigated, 220 hectares were pasture, and 170 were steep hills.

9. From the Hacienda San Francisco, Altepexi received 378 hectares (324 for irrigation and 54 for *agostadero cerril*); from the Hacienda Santa Cruz 592 hectares (64 for irrigation, 248 unirrigated, and 280 *agostadero cerril*); from the Hacienda Venta Negra, 1,202 hectares (1,002 unirrigated and 200 *agostadero cerril*); and from the Hacienda San Andrés, 1,535 hectares of *agostadero cerril.*

10. The payments due at the end of each harvest were high, and the Indians had difficulty raising the money. The owner of the hacienda periodically cut off the water channels, diverting the water to the hacienda. Ultimately, the hacienda owner was able to harass the Indians into selling their water rights to him.

5. *Cooperation and Differentiation*

1. As we have pointed out in our discussion of the history of the valley, a range of cooperative organizations have played a role there, including agrarian unions, cooperatives, political coalitions, and ejidos.

2. Although we do not know what the ownership distributions are in other major communities, our data suggest similar patterns of unequal control over water and, to a lesser degree, land resources.

3. The first thirty-day cycle begins on January 1 at 12:01 A.M. and ends at midnight on January 30. The second cycle begins at 12:01 A.M. on January 31 and ends at midnight on March 1. The third cycle goes from 12:01 A.M. of March 2 until midnight of March 31. The subsequent nine monthly cycles go from the first of the month until midnight on the thirtieth. The five twenty-four hour periods on the thirty-first of May, July, August, October, and December, and February 29 in a leap year are not part of the thirty-day distribution cycle and are distributed as separate *acciones*. The total number of irrigation hours is 8,760 in a nonleap year and 8,784 in a leap year. A *socio*'s *tanda* is the number of continuous hours that he or she is entitled to, and *socios* may have many *tandas* in a monthly cycle. All the associations keep a list of *tandas*, called the *lista de tandeo*, so that each member knows to the minute when his or her *tandas* begin and end.

4. For a study of political patronage and bureaucracy in the Secretaría de Recursos Hidráulicos, see Greenberg 1969 and 1970.

5. These rights and obligations are as follows:

a. Administer water distribution according to the rules.
b. Guard water sources and maintain water works.
c. In compliance with the Labor Laws (Código del Trabajo), hire and fire personnel employed by the *junta*.
d. Keep financial records.
e. Collect and manage users' fees.
f. Inform the ministry when new *junta* personnel are elected and take office.
g. Submit all plans for approval by the ministry and the users.
h. Submit monthly financial statements to the ministry.
i. Call meetings to elect a new *junta de aguas*.
j. Give all requested information to the ministry or its agents, inspector, and employees, and to those who determine the regulations for use.

6. The following are the rules and regulations as they appear in the handwritten records of La Purísima No. 3 (P3):

(a) The association will hold regular meetings of the entire membership on the first Sunday of every month. These meetings will begin at 8:00 A.M., and end by 1:00 P.M. Emergency meetings will be a forum for the discussion of financial affairs, and decisions will be made as to the amount of money that the shareholders must contribute toward the association's operation. Reports will be presented on the association's financial affairs, the operation of the *galerías*, dams, and canals. A report will also be made on the exact distances dug by the *galería* construction and maintenance workers in order to determine the amount of money they will be paid. (b) The governing officers [*mesa directiva*] will consist of a president, secretary, treasurer, and the first, second, and third board members. The officers are to be elected by a majority vote of the association membership for a one-year term. (c) The duties of the officers are to direct the work of the laborers, recruit new laborers, and administer the payroll. The officers will also represent the association in relations with other irrigation associations and organizations. (d) The board members must go as a group to the work sites at the beginning and end of the week to make an exact measurement of the work that has been done. The reason is to calculate how much will be paid and to see how the work is progressing. If this is not done, each board member must, within eight days, pay a fine of four pesos to the association treasury. (e) The failure of a shareholder to attend a meeting that involves proposals by the governing officers will incur a fine of three pesos. (f) Tardiness of any kind by any association member will result in a three-peso fine. Revenue from these fines will be used to purchase tools. (g) Continued tardiness by any association member will result in a 50 percent increase of the fine. (h) A shareholder who fails to make eight consecutive payments of any type will be expelled from the association and will lose the right to all of his or her shares of water. (i) All the shareholders on the association list must stand, if called upon, as candidates for any of the positions of governing officers or board membership. Members who only have one-half of a share are also obliged to stand for office. (j) Any shareholder who refuses to be a candidate for office will be fined ten pesos and will be obliged to take part in association officeholding. (k) If a shareholder is derelict in performing his or her duties as an association officer for a period of at least thirty days, he or she will be expelled by the general assembly and lose shares without any compensation. (l) Any shareholder who is sick will have permission not to attend meetings for sixty days and during this time will be exempt from all association duties. He or she may, however, at any time during these sixty days liquidate his or her holdings and responsibilities without incurring a fine. (m) If a sick member is unable to continue activities in the association, he or she can transfer his or her shares to the association for the same price paid for them. If the association does not want to purchase the share or shares, then the shareholder has permission to transfer the shares to a landholder outside the *municipio* of Altepexi. (n) Any person in the *municipio* who acquires new or additional shares in the association should submit a written report of the purchase to the society and will then be charged twenty-four pesos, which will be the basis for one share or six hours of water. The money will be put in the treasury and used for *galería* construction and maintenance. (o) The association will keep the following books: one for the minutes of both general and emergency meetings, one for accounting, and one for the registry of shareholders. [A book of fines was added in 1950.] (p) When a member dies, his or her sons will inherit all of his or her shares. The share transfer must be confirmed in writing and presented at a general meeting of the association. (q) Neither at work for the association, nor at meetings will a shareholder be permitted to be drunk. A fine will be levied according to the rules covering absences as described above. (r) In *galería* cleaning work, or in general or emergency meetings, minors under eighteen years old will not be permitted to attend in place of adult family members. (s) Any shareholder who is assigned to inform other shareholders of a decision by the board or the governing officers and who fails to do so, will be fined according to the seriousness of the offense.

The positions of president, secretary, and treasurer, the most important and visible association offices, were held by a total of forty members between 1946 and 1973. They occupied eighty-eight positions. Twenty-two, or 55 percent, of the officeholders were *socios mayores*, who held seven (or 8 percent) of the positions in the group. These figures do not indicate that the *socios mayores* held these positions more frequently than the other members. The distribution of the most important administrative positions was about the same as the overall even distribution of all association positions.

The distribution of the board member positions was also even. Of 104 positions, 59 (or 57 percent) were held by 28 of the 53 *socios mayores*. These positions functioned primarily to oversee the actions of the *mesa directiva* and required a fair amount of time; it was not a popular job, but the members served when called on. The special commissions were small temporary groups of three to six members who had a specific duty or function. Upon termination the commission was dissolved.

There were a total of 161 recorded commission positions in P3 over the twenty-eight year period. All of the individuals who participated in association government were also, at one time or another, commission members. The *socios mayores* held 96 (or 60 percent) of the positions, and this represented thirty-seven (or 52 percent) of the seventy-one members who were, at one time or another, commission members. These figures show a slight increase in the participation of *socios mayores* in comparison with the other association positions. If there was a strong relationship between the number of shares held and the participation in association management, the participation of the *socios mayores* was not as high as might have been expected.

When we separated the commission memberships into crisis commissions and routine commissions, a significant distinction became apparent. The membership in routine commissions was more or less the same as the pattern observed for the other positions. Of 122 routine positions, 59 (or 48 percent) were held by *socios mayores*, which indicates that there was actually a slight decrease in participation when compared with the other positions. The crisis commissions, on the other hand, showed a totally different distribution. Of 39 positions, 37 (or 95 percent) were held by *socios mayores*, and 22 out of 24 members were *socios mayores*.

6. *Elites and Irrigation Association Management*

1. The *sociedad civil* is a form of legal incorporation used by small groups. It is similar to the *sociedad anónima* used by commercial enterprises.

BIBLIOGRAPHY

Acheson, James M.
1980 Agricultural Business Choices in a Mexican Village. In *Agricultural Decision Making*, ed. Peggy F. Barlett, pp. 241–264. New York: Academic Press.

Adams, Richard N.
1988 *The Eighth Day: Social Evolution as the Self-Organization of Energy*. Austin: University of Texas Press.

Adams, Robert McC.
1966 *The Evolution of Urban Society: Early Mesopotamia and Prehistoric Mexico*. Chicago: Aldine.

Agribusiness Associates, Inc.
n.d. The Poultry Breeding Industry and Mexican Development. Wellesley Hills, Mass.: unpublished report.

Alexander, Jennifer, and Paul Alexander
1978 Sugar, Rice and Irrigation in Colonial Java. *Ethnohistory* 25(3): 207–223.

Bailey, John J., and John E. Link
1981 Statecraft and Agriculture in Mexico, 1980–1982: Domestic and Foreign Policy Considerations in the Making of Mexican Agricultural Policy. Working Paper in U.S.-Mexico Studies, no. 23. Program in U.S.-Mexico Studies, University of California, San Diego.

Banfield, Edward C.
1958 *The Moral Basis of a Backward Society*. Glencoe, Ill.: The Free Press.

Barkin, David, and Billie R. DeWalt
1988 Sorghum and the Mexican Food Crisis. *Latin American Research Review* 23(3): 30–59.

Barkin, David, and Timothy King
1970 *Regional Economic Development: The River Basin Approach in Mexico*. New York: Cambridge University Press.

Barlett, Peggy F.
1977 The Structure of Decision Making in Paso. *American Ethnologist* 4(2): 285–308.
1980 *Agricultural Decision Making: Anthropological Contributions to Rural Development*. New York: Academic Press.
Barnett, Tony
1984 Small-Scale Irrigation in Sub-Saharan Africa: Sparse Lessons, Big Problems, Any Solutions? *Public Administration and Development* 4:21–47.
Bauer, Dan, and John Hinnant
1980 Revolutionary and Normative Divination. In *Explorations in African Systems of Thought*, ed. Ivan Karp and Charles Byrd, pp. 183–212. Bloomington: University of Indiana Press.
Biswas, Asit K.
1986 Irrigation in Africa. *Land Use Policy* 3 (October): 269–285.
Boerma, A. H.
1975 The World Could Be Fed. *Journal of Soil and Water Conservation* 30 (January/February): 4–11.
Boserup, Esther
1965 *The Conditions of Agricultural Growth*. Chicago: Aldine.
Brandenburg, Frank Ralph
1955 Mexico: An Experiment in One-Party Democracy. Ph.D. dissertation, University of Pennsylvania.
Bratton, Michael
n.d. A Share of the Plow: Small Farmer Organizations and Food Production in Zimbabwe. Unpublished paper.
Bromley, Daniel W., Donald C. Taylor, and Donald E. Parker
1980 Water Reform and Economic Development: Institutional Aspects of Water Management in the Developing Countries. *Economic Development and Cultural Change* 13:365–387.
Brunet, J.
1967 Geologic Studies. In *The Prehistory of the Tehuacan Valley*, Vol. 1, *Environment and Subsistence*, ed. Douglas S. Byers, pp. 66–90. Austin: University of Texas Press.
Butzer, Karl W., Juan F. Mateu, Elizabeth Butzer, and Pavel Kraus
1985 Irrigation Agrosystems in Eastern Spain: Roman or Islamic Origins? *Annals of the Association of American Geographers*, December: 479–509.
Byers, Douglas S.
1967*a* Climate and Hydrology. In *The Prehistory of the Tehuacan Valley*, Vol. 1, *Environment and Subsistence*, ed. idem, pp. 48–65. Austin: University of Texas Press.
1967*b* The Region and Its People. In *The Prehistory of the Tehuacan Valley*, Vol. 1, *Environment and Subsistence*, ed. idem, pp. 34–47. Austin: University of Texas Press.
Cantor, Leonard M.
1970 *A World Geography of Irrigation*. New York: Praeger.

Carlos, Manuel L.
1981 State Policies, State Penetration, and Ecology: A Comparative Analysis of Uneven Development and Underdevelopment in Mexico's Mico Agrarian Region. Working Paper in U.S.-Mexican Studies, no. 19. Program in U.S.-Mexico Studies, University of California, San Diego.

Carta de la Agrupación Agraria de San Gabriel Chilac 21 de julio
1932 *San Gabriel Chilac.* Mexico City: Oficinas Municipales.

Caso, Alfonso
1963 Land Tenure among the Ancient Mexicans. *American Anthropologist* 65:863–868.

Chibnick, Michael
1980 The Statistical Behavior Approach: The Choice between Wage Labor and Cash Cropping in Rural Belize. In *Agricultural Decision Making,* ed. Peggy F. Barlett, pp. 87–114. New York: Academic Press.

Childe, V. Gordon
1936 *Man Makes Himself.* New York: Mentor Books (1951 ed.).

Cleek, Richard K.
1972 The Infiltration Gallery: A Middle Eastern Irrigation System in Southern Mexico. M.A. thesis, University of Texas at Austin.

Climo, Jacob
1978 Collective Farming in Northern and Southern Yucatán, Mexico: Ecological and Administrative Determinants of Success and Failure. *American Ethnologist* 5(2):191–205.

Cockcroft, James D.
1983 *Mexico: Class Formation, Capital Accumulation, and the State.* New York: Monthly Review Press.

Código Agrario
1934 Mexico City: Editorial Porrúa.

Código de Trabajo
1950 Mexico City: Editores Mexicanos Unidos.

Collier, George A.
1974 *Fields of the Tzotzil.* Austin: University of Texas Press.

Comisión del Papaloapan (*see also* Secretaría de Recursos Hidráulicos)
1965 Programa de desarrollo del Valle de Tehuacán. Unpublished report.
1983 Unpublished report.

Conklin, Harold C.
1980 *Ethnographic Atlas of Ifugao.* New Haven: Yale University Press.

Constitución mexicana
1980 Mexico City: Editores Mexicanos Unidos.

Corbett, Jack, and Scott Whiteford
1983 State Penetration and Development in Mesoamerica. In *Heritage of Conquest Thirty Years Later,* ed. Carl Kendall, John Hawkins, and Laurel Bossen, pp. 9–33. Albuquerque: University of New Mexico Press.

Coward, E. Walter, Jr.
1979 Principles of Social Organization in an Indigenous Irrigation System. *Human Organization* 38(1): 28–36.
1980 *Irrigation and Agricultural Development in Asia: Perspectives from the Social Sciences*. Ithaca, N.Y.: Cornell University Press.

Cressey, George B.
1958 Qanats, Karez and Foggaras. *Geographical Review* 48: 27–44.

Cumberland, Charles C.
1952 *The Mexican Revolution: Genesis under Madero*. Austin: University of Texas Press.

Decreto de abril de 1954
1954 San Francisco Altepexi, Puebla, Mexico: Oficinas Municipales.

Decreto del 6 de enero de 1915
1915 Mexico City: Porrúa.

de Janvry, Alain
1981 *The Agrarian Question and Reformism in Latin America*. Baltimore: Johns Hopkins University Press.

de la Peña, Guillermo
1981 *The Legacy of Promises: Agriculture, Politics, and Ritual in the Morelos Highlands of Mexico*. Austin: University of Texas Press.

de los Angeles Crummet, María
1985 Rural Class Structure in Mexico: New Developments, New Perspectives. Kellogg Working Paper, no. 41. Notre Dame, Ind.: Helen Kellogg Institute for International Studies, University of Notre Dame.

Dennis, Phillip A.
1987 *Intervillage Conflict in Oaxaca*. New Brunswick, N.J.: Rutgers University Press.

DeWalt, Billie R.
1979 *Modernization in a Mexican Ejido*. New York: Cambridge University Press.

De Yong, Tim
1981 Searching for the Milagro Beanfield: The Politics of Surface Water in New Mexico. *Public Service* 8(1): 1–8.

Downing, Theodore E.
1974 Irrigation and Moisture-Sensitive Periods: A Zapotec Case. In *Irrigation's Impact on Society*, ed. T. Downing and M. Gibson, pp. 113–122. Tucson: University of Arizona Press.

Dunbar Ortiz, Roxanne
1980 *Roots of Resistance: Land Tenure in New Mexico: 1960–1980*. Los Angeles: Chicano Research Center, University of California.

El Día
1977 Mexico City: 5529.2.

Enge, Kjell I.
1973 The Social Aspects of Irrigation among Mexican Peasant Agriculturalists. American Anthropological Association, 71st Annual Meeting, New Orleans.

1977 Los intereses monetarios de los campesinos: Producción y comercio. *América Indígena* 37(4):1019–1042.

1982*a* Conflict and Control in a Mexican Irrigation System. American Anthropological Association, 81st Annual Meeting, Washington, D.C.

1982*b* Control of Small Irrigation Systems, a Rural Elite and State Institutions: A Mexican Case. Ph.D. dissertation, Boston University.

1983 Irrigation and Conflict: Competition over Resources from an Ecological Perspective. Symposium on Comparative Studies of Irrigation Organization. American Anthropological Association, 82nd Annual Meeting, Chicago, Ill.

1984 Agricultural Development in the Tehuacán Valley, Mexico: Traditional Technologies and New Cooperative Organizations. Symposium on Farming Systems Research: The Integration of Social and Biological Approaches. American Anthropological Association, 83rd Annual Meeting, Denver, Colo.

1985 Irrigation Organizations in Environmental, Social and Political Contexts. Symposium on Socionatural Regions, System Levels and Linkages. Organized by the Anthropology Study Group on Agrarian Systems. American Anthropological Association, 84th Annual Meeting, Washington, D.C.

Enge, Kjell I., and Scott Whiteford

1988 Ecology, Irrigation, and the State in the Tehuacán Valley, Mexico. In *Human Systems Ecology: Studies in the Interaction of Political Economy, Adaptation, and Socionatural Regions*, ed. Sheldon Smith and Ed Reeves. Boulder, Colo.: Westview.

English, Paul Ward

1966 *City and Village in Iran: Settlement and Economy in the Kirman Basin.* Madison: University of Wisconsin Press.

1968 The Origin and Spread of Qanats in the Old World. *Proceedings of the American Philosophical Society* 112:170–181.

Erasmus, Charles J.

1961 *Man Takes Control.* Indianapolis: Bobbs-Merrill.

Esteva, Gustavo

1983 *The Struggle for Rural Mexico.* South Hadley, Mass.: Bergin and Garvey Publishers.

Eyre, John D.

1955 Water Control in a Japanese Irrigation System. *Geographical Review* 45:197–216.

Farrington, I. S., and C. C. Park

1978 Hydraulic Engineering and Irrigation Agriculture in the Moche Valley, Peru, c. A.D. 1250–1532. *Journal of Archaeological Science* 5:255–268.

Fernea, Robert A.

1959 Irrigation and Social Organization among the El Shabāna: A Group of Tribal Cultivators in Southern Iraq. Ph.D. dissertation, University of Chicago.

1963 Conflict in Irrigation. *Comparative Studies in Society and History* 6:76–83.

1970 *Shaykh and Effendi: Changing Patterns of Authority among the El Shaban of Southern Iraq*. Cambridge, Mass.: Harvard University Press.

Finkler, Kaja

1973 A Comparative Study of the Economy of Two Village Communities in Mexico with Special Reference to the Role of Irrigation. Ph.D. dissertation, City University of New York.

1978 From Sharecroppers to Entrepreneurs: Peasant Houshold Production Strategies under the Ejido System of Mexico. *Economic Development and Cultural Change* 27:103–120.

1980 Agrarian Reform and Economic Development: When Is a Landlord a Client and a Sharecropper His Patron? In *Agricultural Decision Making*, ed. Peggy F. Barlett, pp. 265–288. New York: Academic Press.

Flannery, Kent V.

1967 The Vertebrate Fauna and Hunting Patterns. In *The Prehistory of the Tehuacan Valley*, Vol. 1, *Environment and Subsistence*, ed. Douglas S. Byers, pp. 132–177. Austin: University of Texas Press.

1983 Precolumbian Farming in the Valleys of Oaxaca, Nochixtlán, Tehuacán, and Cuicatlán: A Comparative Study. In *The Cloud People: Divergent Evolution of the Zapotec and Mixtec Civilizations*, ed. Kent Flannery and Joyce Marcus. New York: Academic Press.

Fleuret, Patrick

1985 The Social Organization of Water Control in the Taita Hills, Kenya. *American Ethnologist* 10:103–116.

Forman, Shepard

1975 *The Brazilian Peasantry*. New York: Columbia University Press.

Foster, George

1967 *Tzintzuntzan: Mexican Peasants in a Changing World*. Boston: Little, Brown and Company.

Freivalds, John

1972 Farm Corporations in Iran: An Alternative to Traditional Agriculture. *Middle East Journal* 26:185–193.

Friedlander, Judith

1975 *Being Indian in Hueyapan: A Study of Forced Identity in Contemporary Mexico*. New York: St. Martin's Press.

Friedrich, Paul.

1986 *The Princes of Naranja*. Austin: University of Texas Press.

Gates, Gary R., and Marilyn Gates

1972 Uncertainty and Development Risk in Pequeña Irrigación Decisions for Peasants in Campeche, Mexico. *Economic Geography* 48(2): 135–152.

Gaventa, John

1980 *Power and Powerlessness: Quiescence and Rebellion in an Appalachian Valley*. Urbana: University of Illinois Press.

Geertz, Clifford
1959 Form and Variation in Balinese Village Structure. *American Anthropologist* 61:991–1012.
Geo-Re
1981 Informe del estudio hidrogeológico de la Cañada Poblano-Oaxaqueña (Valle de Tehuacán), Estados de Puebla y Oaxaca. Mexico City: Dirección General de Obras Hidráulicas e Ingeniería Agrícola para el Desarrollo Rural, Secretaría de Agricultura y Recursos Hidráulicos.
Gibson, Charles
1964 *The Aztecs under Spanish Rule: A History of the Indians of the Valley of Mexico, 1519–1810*. Stanford: Stanford University Press.
Gil Huerta, Gorgonio
1972 History of the Foundation of the Town of San Gabriel Chilacatla. In *The Prehistory of the Tehuacan Valley*, Vol. 4, *Chronology and Irrigation*, ed. Richard S. MacNeish and Frederick Johnson, pp. 154–161. Austin: University of Texas Press.
Gladwin, Christina
1979 Production Functions and Decision Models: Complementary Models. *American Ethnologist* 6(4):653–674.
Glick, Thomas
1970 *Irrigation and Society in Medieval Valencia*. Cambridge, Mass.: Belknap Press of Harvard University Press.
1972 *The Old World Background of the Irrigation System of San Antonio, Texas*. Southwestern Studies, Monograph 15. El Paso: Texas Western Press.
Goblot, Henri
1979 *Les Qanats: Un technique d'acquisition de l'eau*. Paris: Mouton.
Gray, Robert F.
1963 *The Sonjo of Tanganyika: An Anthropological Study of an Irrigation Based Society*. London: Oxford University Press.
Greenberg, Martin H.
1969 Bureaucracy in Transition: A Mexican Case Study. Ph.D. dissertation, University of Connecticut.
1970 *Bureaucracy and Development: A Mexican Case Study*. Lexington, Mass.: Heath.
Gritzner, Jeffrey A.
1977 *The Origin, Distribution, and Context of Subterranean Aqueducts in Pre-Achaemenid Antiquity: An Exploratory Essay*. Chicago: Department of Geography, University of Chicago.
Guillet, David
1987 Terracing and Irrigation in the Peruvian Highlands. *Current Anthropology* 28:409–430.
Hamilton, Nora
1982 *The Limits of State Autonomy: Post Revolutionary Mexican History*. Princeton: Princeton University Press.
Hanke, Lewis (ed.)
1974 *History of Latin American Civilization: Sources and Interpretations*. Boston: Little, Brown and Company.

Hansen, Roger D.
1971 *The Politics of Mexican Development*. Baltimore: Johns Hopkins University Press.
Hardy, Clarisa
1984 *El estado y los campesinos: La Confederación Nacional Campesina (CNC)*. Mexico City: Editorial Nueva Imagen.
Hartley, John A.
1961 The Political Organization of an Arab Tribe of the Hadhramant. Ph.D. dissertation, London School of Economics.
Henao, Luis Emilio
1978 Las organizaciones campesinas en el Valle de Tehuacán: La organización social de la irrigación. Thesis, Universidad Iberoamericana (Mexico City).
1980 *Tehuacán: Campesinado e irrigación*. Mexico City: Editorial Edicol.
Hernández, Pedro Félix
1965 An Analysis of Social Power in Five Mexican Ejidos. Ph.D. dissertation, Iowa State University of Science and Technology.
Hewitt de Alcántara, Cynthia
1984 *Anthropological Perspectives on Rural Mexico*. London: Routledge and Kegan Paul.
Hoebel, E. Adamson
1966 *Anthropology: The Study of Man*. New York: McGraw-Hill.
Hogg, Richard
1983 Irrigation Agriculture and Pastoral Development: A Lesson from Kenya. *Development and Change* 14:577–591.
Honigmann, John J.
1968 Interpersonal Relations in Atomistic Communities. *Human Organization* 27(3):220–229.
Humlum, J.
1955 Underjordiske vandingskanaler: Kareze, qanat, foggara. *Kulturgeografi* 90:81–131.
Hunt, Eva
1972 Irrigation and the Socio-Political Organization of Cuicatec Cacicazgos. In *The Prehistory of the Tehuacan Valley*, Vol. 4, *Chronology and Irrigation*, ed. Richard S. MacNeish and Frederick Johnson, pp. 162–259. Austin: University of Texas Press.
Hunt, Eva, and Robert Hunt
1974 Irrigation Conflict and Politics: A Mexican Case. In *Irrigation's Impact on Society*, ed. T. Downing and M. Gibson, pp. 129–157. Tucson: University of Arizona Press.
1976 Canal Irrigation and Local Social Organization. *Current Anthropology* 17:389–411.
Hunt, Robert C.
1980 Water Work: Community and Centralization in Canal Irrigation. Unpublished.

Kelly, William W.
1983 Concepts in the Anthropological Study of Irrigation. *American Anthropologist* 85:880–886.
Kirchhoff, Paul
1952 Mesoamerica: Its Geographic Limits, Ethnic Composition, Cultural Characteristics. In *The Heritage of Conquest: Ethnology of Middle America*, ed. Sol Tax. Glencoe, Ill.: Free Press.
Kirk, Rodney Carlos
1975 San Antonio, Yucatan: From Henequen Hacienda to Plantation Ejido. Ph.D. dissertation, Michigan State University, East Lansing.
Kirkby, Anne V. T.
1973 *The Use of Land and Water Resources in the Past and Present in the Valley of Oaxaca, Mexico.* Memoirs of the Museum of Anthropology, no. 5. Ann Arbor: University of Michigan Press.
Lansing, J. Stephen
1987 Balinese "Water Temples" and the Management of Irrigation. *American Anthropologist* 89(2):326–341.
Leach, Edmund R.
1961 *Pul Eliya.* New York: Cambridge University Press.
Lees, Susan H.
1970 Socio-Political Aspects of Canal Irrigation in the Valley of Oaxaca, Mexico. Ph.D. dissertation, University of Michigan.
1974 The State's Use of Irrigation in Changing Peasant Society. In *Irrigation's Impact on Society*, ed. T. Downing and M. Gibson. Tucson: University of Arizona Press.
1976 Hydraulic Development and Political Response in the Valley of Oaxaca, Mexico. *Anthropological Quarterly* 49:107–210.
1986 Coping with Bureaucracy: Survival Strategies in Irrigated Agriculture. *American Anthropologist* 88:610–622.
Leyes y Códigos de México
1981*a* *Ley Federal de Aguas.* Mexico City: Editorial Porrúa.
1981*b* *Ley Federal de Reforma Agraria.* Mexico City: Editorial Porrúa.
Ley Federal de Reforma Agraria, Ley de Fomento Agropecuario
1983 Mexico City: Librerías Teocalli.
López Zamora, Emilio
1968 *El agua, la tierra, los hombres de México.* Mexico City: Fondo de Cultura Económica.
Maass, Arthur, and Raymond Anderson
1978 *. . . and the Desert Shall Rejoice: Conflict, Growth, and Justice in Arid Environments.* Cambridge, Mass.: MIT Press.
MacNeish, Richard S.
1972 Summary of the Cultural Sequence and Its Implications in the Tehuacan Valley. In *The Prehistory of the Tehuacan Valley*, Vol. 5, *Excavations and Reconnaissance*, ed. idem, pp. 496–504. Austin: University of Texas Press.

MacNeish, Richard S., and Antoinette Nelken-Terner

1972 Introduction. In *The Prehistory of the Tehuacan Valley,* Vol. 5, *Excavations and Reconnaissance,* ed. Richard S. MacNeish, pp. 3–13. Austin: University of Texas Press.

Mares, David R.

1980 Articulación nacional-local en el desarollo rural: La irrigación. *América Indígena* 40(3):471–497.

Martínez, Rodolfo

1968 The Mexican Federal System: Its Operation and Significance. Ph.D. dissertation, University of Utah.

Masao, Fidelis T.

1952 The Irrigation System in Uchagga: An Ethno-Historical Approach. *Tanganyika Notes and Records* 75:1–8.

Merrey, Douglas J.

1983 Irrigation, Poverty and Social Change in A Village of Pakistani Punjab: An Historical and Cultural Ecological Analysis. Ph.D. dissertation, University of Pennsylvania.

Millon, René

1954 Irrigation at Teotihuacan. *American Antiquity* 20:177–180.

1957 Irrigation Systems in the Valley of Teotihuacan. *American Antiquity* 23:160–166.

1962 Variations in Social Response to the Practice of Irrigated Agriculture. In *Civilizations in Desert Lands,* ed. Richard B. Woodbury, pp. 56–88. Anthropological Papers, no. 62. Provo: Department of Anthropology, University of Utah.

Millon, René, Clara Hall, and May Díaz

1962 Conflict in the Modern Teotihuacan Irrigation Systems. *Comparative Studies in Society and History* 4:494–524.

Mitchell, William

1973 The Hydraulic Hypothesis: A Reappraisal. *Current Anthropology* 14:532–534.

Orive Alba, Adolfo

1960 *La política de irrigación en México.* Mexico City: Fondo de Cultura Económica.

1970 *La irrigación en México.* Mexico City: Editorial Grijalba.

Padgett, Leon Vincent

1955 Popular Participation in the Mexican "One Party" System. Ph.D. dissertation, Northwestern University, Evanston.

1966 *The Mexican Political System.* Boston: Houghton Mifflin.

Palerm, Angel

1955 The Agricultural Bases of Urban Civilization Mesoamerica. In *Irrigation Civilizations: A Comparative Study,* ed. Julian Steward, pp. 22–42. Washington, D.C.: Pan American Union.

Palerm, Angel, and Eric Wolf

1957 *Ecological Potential and Cultural Development in Mesoamerica.* Studies in Human Ecology, Social Science Monograph 3. Washington, D.C.: Pan American Union.

Paredes Colín, Joaquín
1953 *Apuntes históricos de Tehuacán*. Mexico City: Costa-Amic Editor.
1960 *El distrito de Tehuacán*. Mexico City: Tipográfica Comercial "Don Bosco."
1977 *Apuntes históricos de Tehuacán*, 3d ed. Mexico City: Costa-Amic Editor.
Park, Chris C.
1983 Water Resources and Irrigation Agriculture in Pre-Hispanic Peru. *Geographical Journal* 149(2):153–166.
Paso y Troncoso, Francisco del (ed.)
1905 *Papeles de la Nueva España (PNE)*. 7 vols. Madrid: Tipográfico Sucesores de Rivadeneyra.
1939–1942 *Epistolario de la Nueva España (ENE)*. 16 vols. Mexico City: Biblioteca Histórica Mexicana de Obras Inéditas.
Poleman, T. T.
1964 *The Papaloapan Project: Agricultural Development in the Mexican Tropics*. Stanford: Stanford University Press.
Price, Barbara J.
1971 Pre-Hispanic Irrigation Agriculture in Nuclear America. *Latin American Research Review* 6:3–60.
Rangeley, W. Robert
1987 Irrigation and Drainage in the World. In *Water and Water Policy in World Food Supplies*, ed. Wayne R. Jordan. College Station: Texas A&M University Press.
Raymond, Nathaniel Curtis
1971 The Impact of Land Reform in the Monocrop Region of Yucatan, Mexico. Ph.D. dissertation, Brandeis University.
Reidinger, Richard B.
1974 Institutional Rationing of Canal Water in Northern India: Conflict between Traditional Patterns and Modern Needs. *Economic Development and Cultural Change* 23(1):79–104.
Reyes Osorio, Sergio, et al.
1974 *Estructura agraria y desarrollo agrícola en México*. Mexico City: Fondo de Cultura Económica.
Robins, J. S., and H. F. Rhoades
1958 *Irrigation of Field Corn in the West*. United States Department of Agriculture, Bulletin no. 440. Washington, D.C.: Government Printing Office.
Rogers, Everett
1969 *Modernization among Peasants—The Impact of Communications*. New York: Holt, Rinehart and Winston.
Romanucci-Ross, Lola
1973 *Conflict, Violence and Morality in a Mexican Village*. Palo Alto, Calif.: Mayfield Publishing Company.
Ronfeldt, David
1973 *Atencingo: The Politics of Agrarian Struggle in a Mexican Ejido*. Stanford, Calif.: Stanford University Press.

Rousel Castro, Alberto
1942 Sistema de irrigación antigua del Río Grande de Nasca. *Revista del Museo Nacional* (Lima) 11:196–202.
Rubel, Arthur, and Harriet Kupferer
1968 Perspectives on the Atomistic-Type Society: Introduction. *Human Organization* 27(3):189–199.
Salinas de Gortari, Carlos
1982 *Political Participation, Public Investment, and Support for the System: A Comparative Study of Rural Communities in Mexico.* Research Report Series, no. 35. San Diego: Center for U.S.-Mexican Studies, University of California.
Salmanzadeh, Cyrus, and Gwyn E. Jones
1981 Transformations in the Agrarian Structure in Southwestern Iran. *Journal of Developing Areas* 15:199–214.
Sanders, Thomas
1974 Population Growth and Resource Management: Planning Mexico's Water Supply. *American Field Staff Reports* 2:3.
Sanders, William T., and Barbara Price
1968 *Mesoamerica: The Evolution of a Civilization.* New York: Random House.
Sanderson, Steven E.
1986 *The Transformation of Mexican Agriculture: International Structure and the Politics of Rural Change.* Princeton: Princeton University Press.
Sanderson, Susan
1984 *Land Reform in Mexico: 1910–1980.* New York: Academic Press.
SARH. *See* Secretaría de Agricultura y Recursos Hidráulicos.
Scherr, Sara J.
1985 *The Oil Syndrome and Agricultural Development: Lessons from Tabasco, Mexico.* New York: Praeger.
Schryer, Frans Jozef
1987 Class Conflict and the Corporate Peasant Community: Disputes over Land in Nahuatl Villages. *Journal of Anthropological Research* 43(2): 99–120.
Scott, Ian
1982 *Urban and Spatial Development in Mexico.* A World Bank Publication. Baltimore: Johns Hopkins University Press.
Scott, James
1985 *Weapons of the Weak Everyday Forms of Peasant Resistance.* New Haven: Yale University Press.
Secretaría de Agricultura y Recursos Hidráulicos (SARH)
1982 Estudio socioeconómico y cultural corredor Poblano-Oaxaqueño region de galerías Valle de Tehuacán, Puebla. Mexico City: Subsecretaría de Infraestructura Hidráulica, Dirección General de Obras Hidráulicas e Ingeniería Agrícola para el Desarrollo Rural, Subdirección de Estudos Específicos.
Secretaría de la Presidencia, Comisión de Administración Pública
1970 *Manual de organización del gobierno federal.* Mexico City.

Secretaría de Programación y Presupuesto
1982a *Anuario estadístico de los Estados Unidos Mexicanos—1980.* Coordinación General de los Servicios Nacionales de Estadística, Geografía e Informática. Mexico City.
1982b *Manual de estadísticas básicas del Estado de Puebla.* 2 vols. Coordinación General de los Servicios Nacionales de Estadística, Geografía e Informática. Mexico City.
Secretaría de Recursos Hidráulicos, Comisión del Papaloapan
1975a *Boletín Hidrométrico* no. 22.
1975b *Plan nacional hidráulico.* Mexico City: Subsecretaría de Planeación.
Seele, Enno
1969 Galerías filtrantes en el área de Acatzingo-Tepeaca, Puebla. *Boletín del Instituto Nacional de Antropología e Historia* 35:3–8.
Service, Elman R.
1962 *Primitive Social Organization: An Evolutionary Perspective.* New York: Random House.
Sherbondy, Jeannette E.
1982 The Canal Systems of Hanan Cuzco. Ph.D. dissertation, University of Illinois at Urbana-Champaign.
Simpson, Eyler N.
1937 *The Ejido: Mexico's Way Out.* Chapel Hill: University of North Carolina Press.
Skold, Melvin D., Shinnawi Abdel Atty El Shinnawi, and M. Lofty Nasr
1984 Irrigation Water Distribution along Branch Canals in Egypt: Economic Effects. *Economic Development and Cultural Change* 32(3): 547–568.
Smith, C. Earle
1965a Agriculture, Tehuacán Valley. *Fieldiana Botany* 31:49–100.
1965b Flora, Tehuacán Valley. *Fieldiana Botany* 31:101–143.
Spalding, Rose J.
1988 Peasants, Politics and Change in Rural Mexico. *Latin American Research Review* 23(1):207–219.
Spielberg, Joseph
1968 Small Village Relations in Guatemala: A Case Study. *Human Organization* 27(3):205–211.
Spooner, Brian
1974a City and River in Iran: Urbanization and Irrigation of the Iranian Plateau. *Iranian Studies* 7:681–713.
1974b Irrigation and Society: The Iranian Plateau. In *Irrigation's Impact on Society,* ed. T. Downing and M. Gibson, pp. 43–57. Tucson: University of Arizona Press.
Stevens, Rayfred L.
1964 The Soils of Middle America and Their Relation to Indian Peoples and Cultures. In *Handbook of Middle American Indians,* Vol. 1, *Natural Environment and Early Cultures,* ed. Robert Wauchope and Robert C. West, pp. 265–315. Austin: University of Texas Press.

Steward, Julian H.
1949 Cultural Causality and Law: A Trial Formulation of the Development of Early Civilizations. *American Anthropologist* 51:1–27.

Strickon, Arnold, and Sidney M. Greenfield
1972 *Structure and Process in Latin America: Patronage, Clientage and Power Systems*. Albuquerque: University of New Mexico Press.

Tannenbaum, Frank
1929 *The Mexican Agrarian Revolution*. New York: Anchor Books.

Thomas, Conrad H.
1968 The Agricultural Landscape of the Tehuacán Valley, Mexico. M.A. thesis, University of Calgary.

Troll, Carl
1963 Quanat-Bewassrung in der Alten und Neuen Welt: Ein Kulturgeographisches und Kulturgeschiehtliehes Problem. *Mitteilungen der Osterreichischen Geographischen Gesellschaft* 3:313–330.

Veerman, Leo
n.d. El ex-distrito de Tehuacán. Mimeograph.

Wade, Robert
1975 Water to the Fields: India's Changing Strategy. *South Asian Review* 8(4):301–321.
1979 The Social Response to Irrigation: An Indian Case Study. *Journal of Development Studies* 16:3–26.
1984 Irrigation Reform in Conditions of Populist Anarchy: An Indian Case. *Journal of Development Economics* 14:285–303.

Wallerstein, Immanuel
1984 *The Politics of the World-Economy*. Cambridge: Cambridge University Press.

Warman, Arturo
1976 *Y venimos a contradecir: Los campesinos de Morelos y el estado nacional*. Mexico City: Ediciones de la Casa Chata.
1980 *Ensayos sobre el campesinado en México*. Mexico City: Editorial Nueva Imagen.

Wells, Miriam, and Jacob Climo
1984 Parallel Process in the World System: Intermediate Agencies and Local Factionalism in the United States and Mexico. *Journal of Development Studies* 20(4):151–170.

Whetten, Nathan
1948 *Rural Mexico*. Chicago: University of Chicago Press.

Whiteford, Scott
1986 Troubled Waters: The Regional Impact of Foreign Investment and State Capital in the Mexcali Valley. In *Regional Impacts of U.S.-Mexican Relations*, ed. Ina Rosenthal-Urey. San Diego: Center for U.S.-Mexican Studies, University of California, San Diego.

Whiteford, Scott, and Luis Emilio Henao
1979 Irrigation, Resource Conflict and Selective Migration. In *Migration across Frontiers: Mexico and the United States*, ed. Robert Kemper and

Fernando Cámera, pp. 25–51. Mesoamerican Series. Albany: State University of New York Press.
1980 Irrigación descentralizada, desarrollo y cambio social. *América Indígena* 40(1):57–72.

Whiteford, Scott, and Laura Montgomery
1985 The Political Economy of Rural Transformation: A Mexican Case. In *Micro and Macro Levels of Analysis in Anthropology*, ed. Billie R. DeWalt and Pertti J. Pelto, pp. 147–164. Boulder: Westview.

Wilken, Gene C.
1979 *Studies of Traditional Resource Management in Traditional Middle American Farming Systems*. Preliminary Report to the National Science Foundation, no. 8, pts. 1 and 2.

Wilkinson, John C.
1977 *Water and Tribal Settlement in South-east Arabia*. Oxford: Oxford University Press.

Willey, Gordon R., Gordon F. Ekholm, and René F. Millon
1964 The Patterns of Farming Life and Civilization. In *Handbook of Middle American Indians*, Vol. 1, *Natural Environment and Early Cultures*, ed. Robert Wauchope and Robert C. West, pp. 446–500. Austin: University of Texas Press.

Wittfogel, Karl A.
1957 *Oriental Despotism: A Comparative Study of Total Power*. New Haven: Yale University Press.
1972 The Hydraulic Approach to Pre-Spanish Mesoamerica. In *The Prehistory of the Tehuacan Valley*, Vol. 4, *Chronology and Irrigation*, ed. Richard S. MacNeish and Frederick Johnson, pp. 59–80. Austin: University of Texas Press.

Woodbury, Richard B., and James A. Neely
1972 Water Control Systems of the Tehuacan Valley. In *The Prehistory of the Tehuacan Valley*, Vol. 4, *Chronology and Irrigation*, ed. Richard S. MacNeish and Frederick Johnson, pp. 81–161. Austin: University of Texas Press.

Wulff, H. E.
1968 The Qanats of Iran. *Scientific American* 218:94–105.

Yates, P. Lamartine
1981 *Mexico's Agricultural Dilemma*. Tucson: University of Arizona Press.

Young, Frank W.
1985 The Informant Survey as a Method for Studying Irrigation Systems. *Journal of African and Asian Studies* 20(1–2):56–71.

INDEX

Lightning Source UK Ltd.
Milton Keynes UK
UKOW03f1930171014

240250UK00001B/24/P